2000 Assessment of the Office of Naval Research's Marine Corps Science and Technology Program

Committee for the Review of ONR's Marine Corps Science and Technology Program
Naval Studies Board
Commission on Physical Sciences, Mathematics, and Applications
National Research Council

NATIONAL ACADEMY PRESS
Washington, D.C.

NOTICE: The project that is the subject of this report was approved by the Governing Board of the National Research Council, whose members are drawn from the councils of the National Academy of Sciences, the National Academy of Engineering, and the Institute of Medicine. The members of the committee responsible for the report were chosen for their special competences and with regard for appropriate balance.

This work was performed under Department of the Navy Contract N00014-96-D-0169/0001 issued by the Office of Naval Research under contract authority NR 201-124. However, the content does not necessarily reflect the position or the policy of the Department of the Navy or the government, and no official endorsement should be inferred.

The United States Government has at least a royalty-free, nonexclusive, and irrevocable license throughout the world for government purposes to publish, translate, reproduce, deliver, perform, and dispose of all or any of this work, and to authorize others so to do.

International Standard Book Number 0-309-07138-0

Copyright 2000 by the National Academy of Sciences. All rights reserved.

Copies available from:

Naval Studies Board
National Research Council
2101 Constitution Avenue, N.W.
Washington, D.C. 20418

Printed in the United States of America

THE NATIONAL ACADEMIES

National Academy of Sciences
National Academy of Engineering
Institute of Medicine
National Research Council

The **National Academy of Sciences** is a private, nonprofit, self-perpetuating society of distinguished scholars engaged in scientific and engineering research, dedicated to the furtherance of science and technology and to their use for the general welfare. Upon the authority of the charter granted to it by the Congress in 1863, the Academy has a mandate that requires it to advise the federal government on scientific and technical matters. Dr. Bruce M. Alberts is president of the National Academy of Sciences.

The **National Academy of Engineering** was established in 1964, under the charter of the National Academy of Sciences, as a parallel organization of outstanding engineers. It is autonomous in its administration and in the selection of its members, sharing with the National Academy of Sciences the responsibility for advising the federal government. The National Academy of Engineering also sponsors engineering programs aimed at meeting national needs, encourages education and research, and recognizes the superior achievements of engineers. Dr. William A. Wulf is president of the National Academy of Engineering.

The **Institute of Medicine** was established in 1970 by the National Academy of Sciences to secure the services of eminent members of appropriate professions in the examination of policy matters pertaining to the health of the public. The Institute acts under the responsibility given to the National Academy of Sciences by its congressional charter to be an adviser to the federal government and, upon its own initiative, to identify issues of medical care, research, and education. Dr. Kenneth I. Shine is president of the Institute of Medicine.

The **National Research Council** was organized by the National Academy of Sciences in 1916 to associate the broad community of science and technology with the Academy's purposes of furthering knowledge and advising the federal government. Functioning in accordance with general policies determined by the Academy, the Council has become the principal operating agency of both the National Academy of Sciences and the National Academy of Engineering in providing services to the government, the public, and the scientific and engineering communities. The Council is administered jointly by both Academies and the Institute of Medicine. Dr. Bruce M. Alberts and Dr. William A. Wulf are chairman and vice chairman, respectively, of the National Research Council.

COMMITTEE FOR THE REVIEW OF ONR'S MARINE CORPS SCIENCE AND TECHNOLOGY PROGRAM

BRUCE WALD, Center for Naval Analyses, *Chair*
ALAN BERMAN, Applied Research Laboratory, Pennsylvania State University
A. DOUGLAS CARMICHAEL, Massachusetts Institute of Technology
SABRINA R. EDLOW, Center for Naval Analyses
BRIG "CHIP" ELLIOTT, BBN Technologies
CHARLES A. FOWLER, C.A. Fowler Associates
RAY "M" FRANKLIN, Port Angeles, Washington
DAVID B. KASSING, The Arroyo Center, RAND
R. KENNETH LOBB, Applied Research Laboratory, Pennsylvania State University
IRWIN MENDELSON, Singer Island, Florida
HERBERT RABIN, University of Maryland
DAVID A. RICHWINE, AFCEA International
CHARLES H. SINEX, Applied Physics Laboratory, Johns Hopkins University
MICHAEL G. SOVEREIGN, Monterey, California
JOSEPH ZEIDNER, Bethesda, Maryland

Staff
CHARLES F. DRAPER, Study Director
SIDNEY G. REED, JR., Consultant

NAVAL STUDIES BOARD

VINCENT VITTO, Charles S. Draper Laboratory, Inc., *Chair*
JOSEPH B. REAGAN, Saratoga, California, *Vice Chair*
DAVID R. HEEBNER, McLean, Virginia, *Past Chair*
ALBERT J. BACIOCCO, JR., The Baciocco Group, Inc.
ARTHUR B. BAGGEROER, Massachusetts Institute of Technology
ALAN BERMAN, Applied Research Laboratory, Pennsylvania State University
NORMAN E. BETAQUE, Logistics Management Institute
JAMES P. BROOKS, Litton/Ingalls Shipbuilding, Inc.
NORVAL L. BROOME, Mitre Corporation
JOHN D. CHRISTIE, Logistics Management Institute
RUTH A. DAVID, Analytic Services, Inc.
PAUL K. DAVIS, RAND and RAND Graduate School of Policy Studies
SEYMOUR J. DEITCHMAN, Chevy Chase, Maryland, *Special Advisor*
DANIEL E. HASTINGS, Massachusetts Institute of Technology
FRANK A. HORRIGAN, Bedford, Massachusetts
RICHARD J. IVANETICH, Institute for Defense Analyses
MIRIAM E. JOHN, Sandia National Laboratories
ANNETTE J. KRYGIEL, Great Falls, Virginia
ROBERT B. OAKLEY, National Defense University
HARRISON SHULL, Monterey, California
JAMES M. SINNETT, The Boeing Company
WILLIAM D. SMITH, Fayetteville, Pennsylvania
PAUL K. VAN RIPER, Williamsburg, Virginia
VERENA S. VOMASTIC, The Aerospace Corporation
BRUCE WALD, Center for Naval Analyses
MITZI M. WERTHEIM, Center for Naval Analyses

Navy Liaison Representatives
RADM RAYMOND C. SMITH, USN, Office of the Chief of Naval Operations, N81
RADM PAUL G. GAFFNEY II, USN, Office of the Chief of Naval Operations, N91
    (through June 7, 2000)
RADM JAY M. COHEN, USN, Office of the Chief of Naval Operations, N91 (as of June 8, 2000)

Marine Corps Liaison Representative
LTGEN JOHN E. RHODES, USMC, Commanding General, Marine Corps Combat Development
    Command (through August 17, 2000)
LTGEN BRUCE B. KNUTSON, JR., USMC, Commanding General, Marine Corps Combat
    Development Command (as of August 18, 2000)

RONALD D. TAYLOR, Director
CHARLES F. DRAPER, Senior Program Officer
MARY G. GORDON, Information Officer
SUSAN G. CAMPBELL, Administrative Assistant

COMMISSION ON PHYSICAL SCIENCES, MATHEMATICS, AND APPLICATIONS

PETER M. BANKS, XR Ventures, LLC, *Co-Chair*
WILLIAM H. PRESS, Los Alamos National Laboratory, *Co-Chair*
WILLIAM F. BALLHAUS, JR., The Aerospace Corporation
SHIRLEY CHIANG, University of California at Davis
MARSHALL H. COHEN, California Institute of Technology
RONALD G. DOUGLAS, Texas A&M University
SAMUEL H. FULLER, Analog Devices, Inc.
MICHAEL F. GOODCHILD, University of California at Santa Barbara
MARTHA P. HAYNES, Cornell University
WESLEY T. HUNTRESS, JR., Carnegie Institution
CAROL M. JANTZEN, Westinghouse Savannah River Company
PAUL G. KAMINSKI, Technovation, Inc.
KENNETH H. KELLER, University of Minnesota
JOHN R. KREICK, Sanders, a Lockheed Martin Company (retired)
MARSHA I. LESTER, University of Pennsylvania
W. CARL LINEBERGER, University of Colorado
DUSA M. McDUFF, State University of New York at Stony Brook
JANET L. NORWOOD, Former Commissioner, U.S. Bureau of Labor Statistics
M. ELISABETH PATÉ-CORNELL, Stanford University
NICHOLAS P. SAMIOS, Brookhaven National Laboratory
ROBERT J. SPINRAD, Xerox PARC (retired)

JAMES F. HINCHMAN, Acting Executive Director

Preface

The mission of the Office of Naval Research (ONR) is to maintain a close relationship with the research and development community to support long-range research, foster discovery, nurture future generations of researchers, produce new technologies that meet known naval requirements, and provide innovations in fields relevant to the future Navy and Marine Corps. Accordingly, ONR supports research activities across a broad range of scientific and engineering disciplines. As one means of ensuring that its investments appropriately address naval priorities and requirements and that its programs are of high scientific and technical quality, ONR requires that each of its departments undergo an annual review (with a detailed focus on about one-third of the reviewed department's programs). The Marine Corps Science and Technology (S&T) program reviewed in this report resides within the Expeditionary Warfare Operations Technology Division (Code 353) of the Naval Expeditionary Warfare S&T Department (Code 35) of ONR.

At the request of ONR, the National Research Council (NRC) established the Committee for the Review of ONR's Marine Corps Science and Technology Program to review and evaluate ONR's Marine Corps S&T program components in the areas of maneuver, firepower, logistics, command and control, and training and education against criteria such as the appropriateness of the investment strategy within the context of Marine Corps priorities and requirements, impact on and relevance to Marine Corps needs, Navy/Marine Corps program integration effectiveness, and scientific and technical quality. The committee was also asked to identify promising basic (6.1), exploratory (6.2), and advanced (6.3) research topics that could be initiated to support the Marine Corps S&T program (Appendix A gives the full terms of reference). At the request of the Head of ONR's Naval Expeditionary Warfare S&T Department (Code 35), the committee also reviewed the Extending the Littoral Battlespace (ELB) advanced concept technology demonstration (ACTD).

The committee met once, May 9-11, 2000, in Washington, D.C., to both gather information and prepare an initial draft report. The 3-day meeting was divided into two parts: the first comprised presentations by and interactions with project managers (and ONR-supported principal investigators) responsible for various program components, and the second was devoted to discussing the issues, developing consensus, and drafting the committee's findings and recommendations. The committee's report represents its consensus views on the issues posed in the charge.

Acknowledgment of Reviewers

This report has been reviewed in draft form by individuals chosen for their diverse perspectives and technical expertise, in accordance with procedures approved by the National Research Council's (NRC's) Report Review Committee. The purpose of this independent review is to provide candid and critical comments that will assist the institution in making the published report as sound as possible and to ensure that the report meets institutional standards for objectivity, evidence, and responsiveness to the study charge. The review comments and draft manuscript remain confidential to protect the integrity of the deliberative process. We wish to thank the following individuals for their review of this report:

Anthony J. DeMaria, DeMaria ElectroOptics Systems,
J. Dexter Fletcher, Institute for Defense Analyses,
James J. Harp, Annandale, Virginia,
David W. McCall, Far Hills, New Jersey,
George S. Sebestyen, Systems Development, LLC,
LtGen Philip D. Shutler, USMC (retired), and
H. Gregory Tornatore, Applied Physics Laboratory, Johns Hopkins University.

Although the reviewers listed above provided many constructive comments and suggestions, they were not asked to endorse the conclusions and recommendations, nor did they see the final draft of the report before its release. The review of this report was overseen by Lee M. Hunt, Alexandria, Virginia, appointed by the Commission on Physical Sciences, Mathematics, and Applications, who was responsible for making certain that an independent examination of this report was carried out in accordance with institutional procedures and that all review comments were carefully considered. Responsibility for the final content of this report rests solely with the authoring committee and the institution.

Contents

EXECUTIVE SUMMARY	1
1 INTRODUCTION	6
Context, 6	
General Observations, 11	
Organization of This Report, 13	
2 MANEUVER	14
Overview, 14	
Programs Reviewed, 15	
Recommendations for New Programs, 19	
Concluding Remarks, 20	
3 FIREPOWER	21
Overview, 21	
Programs Reviewed, 22	
Recommendations for New Programs, 25	
Concluding Remarks, 26	
4 LOGISTICS	28
Overview, 28	
Programs Reviewed, 29	
Recommendations for New Programs, 33	
Concluding Remarks, 34	

5 TRAINING AND EDUCATION 35
   Overview, 35
   Programs Reviewed, 36
   Recommendations for New Programs, 40
   Concluding Remarks, 41

6 COMMAND AND CONTROL 43
   Overview, 43
   Programs Reviewed, 44
   Recommendations for New Programs, 52
   Concluding Remarks, 54

7 BASIC RESEARCH (6.1) 55
   Overview, 55
   Programs Reviewed, 55
   Recommendations for New Programs, 62
   Concluding Remarks, 63

8 EXTENDING THE LITTORAL BATTLESPACE (ELB) ADVANCED CONCEPT
   TECHNOLOGY DEMONSTRATION 65
   Findings, 66
   Recommendations, 67
   Concluding Remarks, 69

9 SUGGESTIONS FOR IMPROVING PROGRAM EFFECTIVENESS AND
   ACHIEVING BETTER INTEGRATION WITH THE MARINE CORPS 70
   The Code 353 Program, 70
   Code 353 and Other Parts of the Office of Naval Research, 72
   The Opportunity, 74

APPENDIXES

A Terms of Reference 77
B Previous Training and Education Studies 78
C Committee Biographies 85
D Acronyms 89

Executive Summary

This review of the Science and Technology (S&T) program of the Office of Naval Research's (ONR's) Expeditionary Warfare Operations Technology Division, Code 353, comes at a time of considerable change in the Marine Corps and in ONR, which are currently in the midst of significant transitions. The Marine Corps is making plans to equip and train for engaging in a new style of warfare known as Operational Maneuver From the Sea (OMFTS)[1] and for performing a wide variety of missions in urban settings, ranging from humanitarian assistance to combat and mixes of these suggested by the term *three-block war*. During 1999, ONR assumed management of that portion of the Marine Corps S&T program that had not been assigned several years earlier to the Marine Corps Warfighting Laboratory (MCWL). In 2002, control of most of ONR's advanced development funding (6.3), and of much of its exploratory development funding (6.2), will move from ONR's line divisions, of which Code 353 is one of many, to 12 new program offices, each dedicated to demonstrating technologies for future naval capabilities (FNCs).

Given these changes, it is not surprising that some of the projects inherited recently by ONR, and assessed by the Committee for the Review of ONR's Marine Corps Science and Technology Program under the auspices of the Naval Studies Board of the National Research Council, differed from the customary ONR project and were more akin to preacquisition or acquisition support than to S&T. It is also not surprising that Code 353 could not articulate its plans for future investments clearly and concisely, given the current uncertainty about the content of and funding level for FNCs.

The Marine Corps S&T program supports the five imperatives for technology advancement that the Marine Corps Combat Development Command (MCCDC) has identified as prerequisites for the transition to OMFTS: maneuver, firepower, logistics, training and education, and command and control. The committee supports investment in these areas and, in the report's discussions and recommendations, follows the five imperatives.

---

[1]Headquarters, U.S. Marine Corps. 1996. "Operational Maneuver From the Sea," U.S. Government Printing Office, Washington, D.C., January 4.

Within recent months, Navy 6.1 (basic research) funds have been made available to Code 353 to initiate a basic research program in technology topics with the potential for applications to support Marine Corps priorities and requirements. Although this program is too new to permit evaluation of its progress, the report comments on the topics selected and proposes additional topics for new investment.

The body of this report describes in detail the committee's findings and recommendations concerning the individual projects now being pursued by Code 353. At the request of the Head of ONR's Naval Expeditionary Warfare S&T Department (Code 35), the committee also reviewed the Extending the Littoral Battlespace (ELB) advanced concept technology demonstration (ACTD).[2] The recommended actions, which include continuation, redirection, consideration for support from other budget categories, and termination, are summarized in Table ES.1. The report also offers recommendations for alternative future Code 353 S&T investments. These recommendations are summarized in Table ES.2.

The strategy recommended for ONR that forms the basis of the committee's individual recommendations on the Code 353 program can be described as follows:

- Eliminate from the Code 353 program, at an orderly but determined pace, preacquisition and other activities that do not conform to the usual ONR S&T standards of innovation and technical aggressiveness.[3]
- Leave system demonstrations principally to MCWL, fleet battle experiments (FBEs), and the FNCs.
- Embark on a discovery program to identify and refine technologies that can have a substantial payoff in achieving OMFTS.
- Exploit the talents and insights of other ONR divisions in program formulation and performer selection.

With regard to the last point, the committee approves of the Code 353 strategy of physically locating its program officers near other ONR program officers working in similar scientific fields. However, although propinquity may be a necessary condition for successful informal interaction, it is far from a sufficient condition. Code 353 will have to make a special effort to achieve effective liaison. The committee recommends that ONR management demonstrate its commitment to supporting this interaction by doing the following:

- Providing additional staff to Code 353 so that program officers can interact effectively with their ONR peers and Code 353 can participate more forcefully in the management of the Marine Corps S&T program, and
- Communicating clearly that support of the Marine Corps is the responsibility of the entire Office of Naval Research.

The committee believes that Code 353 has other important work to do beyond formulating and managing the execution of the Marine Corps S&T program and the supporting 6.1 program. Examples follow below.

---

[2]The ONR ELB ACTD does not fall under the purview of ONR Code 353; rather, it is a program office reporting directly to ONR Code 35.

[3]ONR's mission is to maintain a close relationship with the research and development community in order to support long-range research, foster discovery, nurture future generations of researchers, produce new technologies that meet known naval requirements, and provide order-of-magnitude innovations in fields relevant to the future Navy and Marine Corps. See ONR's description online at <http://www.onr.navy.mil/sci_tech>.

TABLE ES.1 Summary of Recommendations for Code 353 Projects and the ELB ACTD

	Project	Recommendation
Imperative for OMFTS		
Maneuver	JMDT	Enlist other institutions in conducting controlled sensor experiments and in gathering, analyzing, and publishing databases.
	RST-V	Consider other power plants such as the micro gas turbine, and evaluate lithium battery safety.
	MEFF-V	Support, but periodically reevaluate payoff.
	Autonomous Operations	Design new vehicles to accommodate autonomy. Monitor the work of the National Aeronautics and Space Administration and others.
	Simulation-based Acquisition	Ensure Marine Corps specificity; include cost. Reexamine funding category.
Firepower	ETAL	Look beyond ETAL to targeting multiple mobile targets.
	OCSW	Maintain a funding line for new starts.
	CLAWS	Transfer to future naval capability.
Logistics	Bulk Liquids	Complete and transition EFS. Continue sensor exploration. Consider water.
	Corrosion	Focus on AAAV.
	Modeling and Simulation	Simplify TLoaDS setup. Assess airlift vulnerability.
	SUL	Use error-correcting codes. Devise means of synchronizing databases. Develop and test decision aids.
Training and Education	SUTT	Quantify payoff. Develop a formal evaluation plan.
	Motion Capture	Continue, but demonstrate cost-effectiveness of fidelity.
	Wearable Computers	Continue through demonstration; include evaluation.
	RSEDG (SBIR Phase II)	Continue through demonstration; include evaluation.
	STO Scoring	Withhold approval until angular accuracy is verified.
	MOUT	Continue through demonstration and evaluation.
	MAGTF FOM	Review, reevaluate feasibility, and adjust as indicated.
	MAGTF XXI and TDGs	Incorporate evaluation plans.
Command and Control	Harsh Environments	Terminate.
	JTRS Mobile Network	Minimize JTRS requirements accretion. Explore ad hoc networking technology.
	JTRS Antenna	Continue funding, but seek alternatives to body mounting.
	Mobile DF	Coordinate with the National Security Agency and others. Focus on vehicle-mounted antenna performance. Reexamine funding category.
	PC-based TDOA	Review feasibility; terminate or redirect to exploration of various processing algorithms.
	Wideband Tactical Communications	Map out experiments. Consider for incorporation in SUL ACTD.

TABLE ES.1 Continued

	Project	Recommendation
6.1	UWB Ranging	Consider for 6.2 funding.
	UWB Channel Coding	Consider for 6.2 funding.
	Low-power CMOS	Consider for 6.3 funding.
	Multiple Sensors	Continue, but structure to account for Marine Corps needs.
	Multisource Mobile	Program review by Code 31.
	Thin-film Batteries	Continue materials task, but transition prototyping to 6.2.
	Fuel Cells	Look for a transition to 6.2.
	Modeling Power Systems	Look for an early transition to 6.2.
ELB	NA	Focus on experimentation.
		Minimize development investment.
		Learn how to accommodate outages.
		Perform a security analysis.
		Quantify capacity requirements.
		Collaborate.

NOTE: For definitions of acronyms, see Appendix D.

Because the OMFTS concept has usually been described qualitatively and without target dates, no mechanism exists for quantitatively assessing the payoff in mission effectiveness that a possible improvement in technical performance would provide, and no guidance is given regarding target dates for meeting technology needs. Without such mechanisms and explicit milestones, ONR will find it hard to transform a largely performer-driven program into one responsive to requirements and opportunities. Without such mechanisms, opportunities to adjust performance goals across the imperatives and to rebalance investments accordingly may be overlooked. For example, qualitative thinking suggests that precision targeting information allowing efficient and effective use of ordnance can reduce the need for logistics capacity, but quantitative analysis is needed to establish the extent to which technologies that provide more accurate targeting can decrease the lift required for munitions. Similarly, the lack of quantitative OMFTS goals—concerning, for example, the size of forces that must maneuver directly from ships to their objective and the distances involved—makes it difficult to assess the adequacy of systems based on current or clearly foreseeable technology.

Because the Marine Corps has no organizational component responsible for establishing S&T requirements—a function performed instead by a working group that meets periodically—the committee recommends that Code 353 take an active role in:

• Urging quantitative analysis of the OMFTS concept and providing, if necessary, some of the support and technology for such analyses; and
• Creating technology roadmaps that predict the performance levels and availability dates of the systems that are based on technologies that Code 353 and others are exploring.

Code 353 program officers should achieve more in their interactions with their peers than an exchange of ideas about how to direct the Marine Corps S&T program. They should make their peers aware of Marine Corps needs so that, to the greatest extent possible, the Navy S&T program—which

TABLE ES.2 Summary of Topics Recommended for New Investments

	Topic
Imperative for OMFTS	
Maneuver	None, other than 6.1 research in mine detection phenomenology
Firepower	Concepts for weapons of controllable lethality
	New concepts for mortars, hand grenades, and rifle grenades
	Countersniper weapons
	Sensor systems to support counterbattery weapons
Logistics	Vulnerability of sea-based logistics train
	Shipment development/item location in shipboard environment
	Innovative approaches to water usage
Training and Education	General military training
	Proficiency training
	Team training
	Limitations of close-combat computer games
	A long-range research and technology plan for overall training and education that includes the rationale for selecting programs
Command and Control	Connectivity to joint and national sensors
	Mobile networking for Marine-specific problems
	Network security issues for MOUT and OMFTS
	Networking the AAAV
	Deception and other information operations
6.1	Investigation of other phenomena that could be incorporated into weapons of controllable lethality
	Other fundamental research that could lead to better devices and techniques for urban warfare
	Phenomena and devices for detecting mines, particularly remotely, and for mine countermeasures
	Techniques for gaining and disseminating situational awareness in a Marine Corps context to enable informed maneuver
	Materials that could reduce the logistics burden of OMFTS

dwarfs the Marine Corps program in funding level—will benefit the Corps. Enhancing awareness of requirements for fulfilling the Marine Corps mission can be accomplished through program-specific interactions and also by facilitation of interchanges between ONR program officers, ONR-supported researchers, and the Marine Corps customers for ONR's technology. Code 353 also should reach out beyond the naval services to learn about other technology developers and to sensitize those developers to Marine Corps needs.

Overall, the shortcomings of the specific projects reviewed by the committee are not overly alarming, given the history of the program and the short time it has been managed by ONR. The committee hopes that ONR and Code 353 will treat the circumstances recounted at the beginning of this summary as an opportunity to formulate a new, technically exciting program that will help the Marine Corps achieve its goals for OMFTS.

1

Introduction

CONTEXT

The Office of Naval Research (ONR) asked the Naval Studies Board (NSB) of the National Research Council to review ONR's Science and Technology (S&T) program that supports the Marine Corps. This review occurred in May 2000 in the context of four ongoing transitions that some might even call revolutions:

- Building a capability for Operational Maneuver From the Sea,
- Preparing for the challenges of the so-called *three-block war*,
- Entrusting management of Marine Corps S&T to ONR and the Marine Corps Warfighting Laboratory, and
- Repackaging much of ONR's S&T activity into 12 future naval capabilities programs.

Each of these transitions poses challenges to the Marine Corps S&T program and also affected the approach taken in this review.

Ongoing Transitions Affecting the S&T Program

Operational Maneuver From the Sea

The Marine Corps, faced with new threats to traditional amphibious operations and to the Navy assets that supported them, plans to equip and train for a new style of expeditionary warfare known as Operational Maneuver From the Sea (OMFTS).[1] Rather than forces taking large beachheads and

---

[1] Headquarters, U.S. Marine Corps. 1996. "Operational Maneuver From the Sea," U.S. Government Printing Office, Washington, D.C., January 4.

engaging in subsequent attrition warfare, this concept emphasizes informed maneuver and envisions highly mobile forces holding large inland areas at risk and then maneuvering rapidly to achieve objectives while avoiding concentrated enemy forces.

Accomplishing OMFTS will require assault platforms of high speed and range that can be launched from Navy platforms at least 25 nautical miles from hostile shores. The Marine Corps has invested heavily in two platforms—the advanced amphibious assault vehicle (AAAV) with high mobility on both sea and land and the vertical take-off and landing V-22 aircraft—to meet some of these requirements, and the Navy is investing in weapons such as the 5-in. gun-fired extended-range guided munition (ERGM) and the land-attack standard missile (LASM), a variant of the missile used in the Navy Aegis air defense system, to provide long-range fire support to maneuvering Marine Corps elements, thereby reducing some of the logistics burden of bringing armor and artillery pieces ashore.

However, completion of the transition to OMFTS will require meeting many other challenges, including the provision of the following:

- Complete and accurate situational awareness to permit light Marine forces to avoid concentrated enemy forces and mined transit lanes,
- In-stride clearance of mines and obstacles to avoid giving the enemy time to mass defensive forces,
- Capabilities for initial assaults over great distances by forces of significant size and power,
- Long-range fire to deliver effective munitions from ship stand-off distances to inland objective areas,
- Equipment and procedures to target this long-range fire,
- Logistics capacity and flexibility to resupply maneuvering forces, and
- Capabilities to defend assault and logistics vehicles from unavoidable enemy fire.

Meeting these challenges will likely require S&T investment.

On the other hand, although OMFTS attempts to avoid major firefights between large forces, these confrontations may occur in the future, particularly when situational awareness is incomplete. The Marine Corps must be prepared for these encounters during and beyond the period of transition to OMFTS and will continue to invest to some extent in the equipment for attrition warfare. The challenges for the S&T community are as follows:

- Balance its investment to enhance the capacity to engage in the old and the new styles of warfare, and
- Coordinate investments in ground warfare with the Army and in air warfare with the Navy and the Air Force.

The Three-Block War

Although the Marine Corps must remain ready to achieve forcible entries in major conflicts between nation states, military operations in support of U.S. national interests will more frequently involve the challenging mix of armed confrontations with irregular forces, peacekeeping, and humanitarian operations that has been dubbed the three-block war.

Many of these operations will take place in areas with dense civilian populations. Military Operations in Urban Terrain (MOUT) were undertaken by the Marines in heavy combat at Seoul, Korea, in 1950 and Hue, Vietnam, in 1968 and more recently in the evacuation of Mogadishu, Somalia, in 1997.

However, MOUT today pose new challenges: not only must they accomplish military objectives but they must also accomplish them in a way that will not undermine the political objectives of the military operations given modern sensitivities.[2]

The Marine Corps should expect frequent involvement in three-block wars because, throughout U.S. history, the nation has called on the Marines for paramilitary activity on foreign shores. However, the Marine Corps and the other Services must wrestle with such challenges as the following:

- Linguistic and cultural diversity hampering intelligence gathering and psychological operations;
- Difficulties in sensing, targeting, and navigating in urban areas;
- Limited mobility in urban areas; and
- Lack of weapons having controllable lethality.

Meeting these challenges will likely require S&T investment, but the Marine Corps must coordinate its investments with those made by other Services, by the Defense Advanced Research Projects Agency, and by the special operations, psychological operations, and intelligence communities.

Management of Marine Corps S&T

The third major transition affecting the committee's review has been the transition in the management of Marine Corps S&T. In the past, the Marine Corps Systems Command (MARCORSYSCOM) managed the Amphibious Warfare Technology (AWT) program. Although the committee did not study the program's history in detail, it believes that the program was closely tied to existing or planned acquisition programs of record. This tie to preacquisition activities and acquisition support ensured transition opportunities for successes in the S&T program but limited that program's technological horizon and constrained its capability in accord with the somewhat ponderous pace of acquisition programs.

Several years ago, recognizing the need for the rapid co-evolution of capabilities and doctrine, the Marine Corps established the Marine Corps Warfighting Laboratory (MCWL) as a component of the Marine Corps Combat Development Command (MCCDC), the major command with responsibility for shaping the future Marine Corps. Although Congress provided incremental Marine Corps S&T funding to establish the MCWL, the S&T total has subsequently declined. Over one-half of the total is committed to the MCWL to support experimentation with warfighting concepts, and the baseline AWT program was left with only about one-half of the funding it had before the establishment of the MCWL. The committee was not asked to review MCWL activities. Figure 1.1, adapted from a slide presented to the committee by Code 353, traces this funding history.

In 1999, the Commander of the MCWL was assigned additional duties as Vice Chief of Naval Research, and management of the remaining portion of the AWT program was transferred to ONR's Code 353, its Expeditionary Warfare Operations Technology Division. That division is led by a civilian with extensive Marine Corps experience and by his Marine Corps officer deputy. Program management has been entrusted to two officers and a civilian; two additional officers are being sought. Much of the determination of specific program directions has apparently been left to the initiative of the performing

---

[2]For additional reading on MOUT, see Marine Corps Combat Development Command. 1997. "A Concept for Future Military Operations on Urbanized Terrain," United States Marine Corps, Quantico, Va., July 25. Available online at <http://www.concepts.quantico.usmc.mil/mout.htm>.

INTRODUCTION

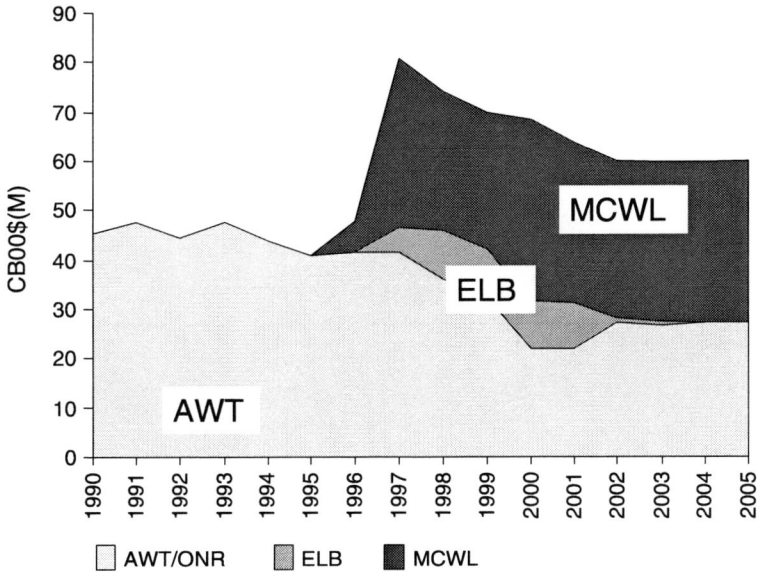

FIGURE 1.1 Profile of funding for Marine Corps S&T, 1990 to 2005. Courtesy of the Office of Naval Research. For definitions of acronyms, see Appendix D.

warfare centers such as the Space and Naval Warfare Systems Command (SPAWAR) Systems Center and the Naval Sea Systems Command Naval Surface Warfare Center (NSWC).

The program managers have been physically dispersed to spaces adjacent to the ONR scientific officers managing similar technologies to encourage interactions between the applications domain expertise of the Code 353 junior staff and the scientific expertise of their new neighbors.

Both the committee and Code 353 recognize that some of the preacquisition and acquisition support work initiated under the MARCORSYSCOM AWT program lacks the technological aggressiveness that characterizes ONR programs and indeed may not even meet the strict definition of S&T, so that some S&T funding is being spent for work that is not really S&T. However, to the extent that they are necessary components of acquisitions that meet Marine Corps needs, these programs cannot be summarily terminated and must be transitioned in an orderly fashion to more appropriate budget categories.

Within FY00, $1 million of Navy 6.1 funding was made available to Code 353 to initiate a basic science program in support of potential Marine Corps needs, and a larger 6.1 effort is contemplated for FY01. At the time of the committee's review, the initial awards from that science program had just been made, so there was no progress to assess.

ONR S&T Management

Although ONR integrated the management of related 6.1 (basic research), 6.2 (exploratory development), and 6.3 (advanced development) activities into the same departments in the mid-1990s to facilitate transition, frustration continued over the relative infrequency with which ONR investments visibly influenced fleet capabilities. Most 6.2 was conducted on a level-of-effort basis, and the process for establishing an advanced technology demonstration (ATD) to exhibit the fruits of exploratory development was too uncertain and took too long.

During 1999, ONR embarked on a new course intended to overcome these difficulties. Beginning in FY02, most of ONR's 6.3 funding and about one-half of its 6.2 funding will be concentrated into 12 programs called future naval capabilities (FNCs). Each FNC is a long-term commitment, typically running through the program objective memorandum (POM) years and beyond, to focus investments in a particular area and to harvest promptly technologies suitable for demonstration, and, by demonstrating them, convince the acquisition community to initiate engineering developments that will lead to new naval capabilities. Each FNC is designed to produce its first demonstration within 2 years of initiation and subsequent demonstrations periodically thereafter. Combined with reforms in defense acquisition designed to encourage spiral development,[3] FNCs should lead to the prompt incorporation of new technology into naval systems.

Most of ONR's 6.1 and about one-half of its 6.2 resources are reserved for sustaining what are now called discovery programs, which will produce technology that will be harvested in future FNCs.

The Committee's Approach

Exciting as the institutional change within ONR may be, and irrespective of its likely eventual benefits to the Department of the Navy, it posed a problem to the committee. This review was chartered by the ONR department that contains Code 353, but the FNCs will not be managed by ONR's line departments but by new program offices. Although work that logically follows current Code 353-sponsored activities has been proposed, it is far from clear what will be performed under the direction of Code 353 and what will be performed under FNC management. FNCs in Warfighter Protection, Organic Mine Countermeasures, Missile Defense, Capable Manpower, Time Critical Strike, and Expeditionary Logistics are likely to subsume some of these activities, but the FNC portfolio has not yet been determined. In a few cases, the committee was advised not to comment extensively on a program that was sure to be folded into an FNC.

Because the scope of funding for the FNCs had not been determined at the time of the review, Code 353 could not propose specific investment levels for its future discovery program.

As a consequence of the transition in the management of Marine Corps S&T and the transition to FNCs in ONR S&T management, the technologically oriented members of the committee found themselves in a difficult position. The MCWL program was not visible to the committee. Some of the old AWT program efforts that had been entrusted to Code 353 presented few technological challenges; some of the contemporary programs were likely to become part of the FNCs and were therefore not within the committee's purview. No clear investment strategy for new discovery programs could be presented because of uncertainty about what would remain within Code 353's charter and how much money would be available to execute that charter.

Anticipating the committee's frustration by these circumstances, ONR's terms of reference requested not only an evaluation of the current program (an evaluation summarized in Table ES.1) but also requested the identification of promising alternative investments. The committee's response to that latter task is summarized in Table ES.2, but it left to others, except in a few obvious cases, the determination of whether some of this work was being or would be pursued at the MCWL or in FNCs.

---

[3]The spiral process, also called evolutionary development of requirements and systems, is an innovative method for fielding a system quickly by using commercial and government off-the-shelf equipment, with maximum user involvement throughout the process (Naval Studies Board, National Research Council. 2000. *Network-Centric Naval Forces: A Transition Strategy for Enhancing Operational Capabilities*, National Academy Press, Washington, D.C., pp. 294-295).

GENERAL OBSERVATIONS

Addressing in subsequent chapters this study's terms of reference, the committee in this section makes a number of general observations offered for consideration by the Marine Corps as it develops its S&T program.

Lack of Quantitative System Analyses[4]

During the development of the OMFTS concept, MCCDC identified the following five "imperative" capabilities to enable OMFTS:

- Maneuver,
- Firepower,
- Logistics,
- Training and education, and
- Command and control.

ONR uses this framework to organize its S&T. Its command and control program also includes work in communications systems.

The committee agrees with this list of imperatives, although it would expand the scope of command and control to include intelligence, reconnaissance, and surveillance—other important elements of information superiority.

Formulating an S&T investment strategy requires knowing not only the prospects for improvements in these areas, but also the degree of improvement needed to make OMFTS work. The committee saw considerable qualitative thought in ONR's description of OMFTS concepts, but, except in logistics, only limited quantitative analysis. The committee did not see any systematic parsing of concepts into required technologies.

For example, although the Navy is developing ERGM to meet a range specification supplied by the Marine Corps, the committee did not see analyses that helped specify needed numbers of weapons or rates of fire as a function of operational situations. The committee heard some ideas about how radios and direction finders should perform but received no information on criteria motivating the desired levels of performance.

Not only should the Marine Corps perform and exhibit the analyses that lead to performance goals within an imperative, but it should also identify possibilities for trade-offs between imperatives. An obvious trade-off is that between precision and related logistic requirements; for example, the more precisely an enemy location is known and the more precisely the weapons can be aimed, the less demanding the requirement is for transporting ammunition. Other trade-offs doubtless exist and should be explored. MCCDC is probably the appropriate activity for this exploration, given its concept development mission and laboratory and simulation capabilities.

---

[4]The lack of quantitative system analyses is a finding of previous Naval Studies Board reports with regard to science and technology planning. See (1) Naval Studies Board, National Research Council. 2000. *Network-Centric Naval Forces: A Transition Strategy for Enhancing Operational Capabilities,* National Academy Press, Washington, D.C.; (2) Naval Studies Board, National Research Council. 2000. *An Assessment of Undersea Weapons Science and Technology,* National Academy Press, Washington, D.C.; and (3) Naval Studies Board, National Research Council. 1999. *1999 Assessment of the Office of Naval Research's Air and Surface Weapons Technology Program,* National Academy Press, Washington, D.C.

The interactions between ONR and MCCDC should be two-way. The OMFTS vision raises challenges whose degree is determined by force size and distances envisioned in the Ship-to-Objective Maneuver (STOM) concept.[5] The complexity associated with executing an OMFTS maneuver can vary by orders of magnitude depending on the combination of force size and STOM distance, not to mention other factors such as whether the environment is hostile or the transit lanes are mined. Technology choices and investment will vary depending on the particular scenario, which the Marine Corps ultimately needs to define.

Therefore, ONR should identify breakthrough and enabling technologies and their projected performance improvements and availability dates. This would guide the Navy and Marine Corps in determining the extent to which OMFTS force size and STOM distance can be increased and in planning time-phased acquisitions. The two-way interaction would provide both focus for the technologists and guidance for the planners.

Relative Neglect of Joint Operations

The committee expected that the program would make many references to network-centric operations[6] in general and particularly to the use of multiple nonorganic sensors to provide the exquisitely detailed and accurate situational awareness needed by small, lightly armored ground elements. However, it found little or no commitment to network-centric operations and few systematic analyses of the trade-offs between using organic capabilities and relying on nonorganic support.

Neglect of Deception and Concealment

OMFTS and STOM envision avoidance of the enemy by light, maneuvering forces. These forces not only must possess excellent situational awareness to avoid blundering into concentrated opposition, but also must keep the adversary from determining their nature and intentions. Although the committee anticipated presentations about technologies that would assist the Marines in concealment from and deception of opposing forces, it did not hear about any programs with this aim.

Relative Neglect of MOUT

Although a few of the projects reviewed were applicable to MOUT and the three-block war, most were not. Urban operations require different capabilities than OMFTS, even though they are more likely than major regional conflicts. Neglect of applicable S&T could lead to a failure to supply Marines with the materiel needed for effective MOUT.

---

[5]Van Riper, LtGen Paul K. (Ret.). 1997. "A Concept for Ship-to-Objective Maneuver," *Marine Corps Gazette,* Marine Corps Association, Quantico, Va., November. The original "Ship-to-Objective Maneuver" Marine Corps concept paper is available online at <http://www.concepts.quantico.usmc.mil/stom.htm>.

[6]Network-centric operations are military operations that exploit state-of-the-art information and networking technology to integrate widely dispersed human decision makers, situational and targeting sensors, and forces and weapons into a highly adaptive, comprehensive system to achieve unprecedented mission effectiveness (Naval Studies Board, National Research Council. 2000. *Network-Centric Naval Forces: A Transition Strategy for Enhancing Operational Capabilities,* National Academy Press, Washington, D.C., p. 1).

Performer-determined Goals

The Marine Corps has no full-time organization analogous to the Navy's N91 that can set S&T goals based on its oversight of the research, development, test, and evaluation budget and the difficulties in meeting operational requirements encountered by system developers. Formerly, Marine Corps S&T goals were set by occasional roundtables of technology suppliers and users; recently, a working group was established that has resolved to meet periodically.

This situation, combined with the lack of quantitative system analysis as discussed above in "Lack of Quantitative System Analyses" (page 11), the relative inexperience of some Code 353 personnel in the management of S&T, and the consequent delegation of technical oversight to the system command warfare centers that execute the program, may have led to a portfolio of projects within which priorities and goals were set by the performers on the basis of their capabilities and desires rather than Marine Corps needs.

ORGANIZATION OF THIS REPORT

Each of the five chapters (Chapters 2 through 6) that follow pertains to one of the five MCCDC imperatives. Each chapter begins with an overview of the imperative, proceeds to the findings and recommendations for each project presented to the committee at its May 2000 meeting, and concludes with suggestions for other S&T investments and general remarks. The next two chapters (Chapters 7 and 8) follow a similar outline for 6.1 programs and for the Extending the Littoral Battlespace (ELB) advanced concept technology demonstration (ACTD), which is not part of Code 353's program and is managed by another division in ONR's Naval Expeditionary Warfare S&T Department.

These chapters are followed by a final chapter (Chapter 9) presenting suggestions for improved program effectiveness and integration with the Marine Corps. Appendix A reproduces the terms of reference under which the committee operated, Appendix B presents highlights from three past training and education studies, Appendix C gives short biographies of the committee members, and Appendix D defines the acronyms used in this report.

2

Maneuver

OVERVIEW

Marine Corps concepts of maneuver rely on information to achieve *informed maneuver*. Maneuver implies dynamic objectives, dynamic threat overlays, and dynamic courses of action planned and executed.

This dynamic approach to warfighting requires the following:

- Timely and cohesive command, control, communications, computing, and intelligence (C4I) that leverage network-centric warfare advances;
- Fundamental changes in logistics approaches and logistic requirements for maneuver elements;
- Creative approaches to transport across the sea, land, and the sea-land interface, with the sea-land interface transport requirements unique to the Marine Corps;
- Forces tailored for the spectrum of conflict, from operations other than war to major theater wars; and
- Timely and cohesive information, leveraging information sources ranging from force reconnaissance units on the ground to unmanned aerial vehicles (UAVs) to national satellite systems.

Information requirements to support maneuver vary with the range of conflict. Conflicts occurring in rural settings require large-area information gathering. Urban environments impose considerably more complex information requirements: they involve a closely gridded, three-dimensional space with highly perishable information and more risks of collateral damage. Information overload is a potential hazard, and the transformation of information into knowledge requires multisource processing, fusion, and conflict resolution. The need for 6.1 research that could enable these capabilities is discussed in Chapter 7 in "Recommendations for New Programs" (page 62).

ONR's Marine Corps maneuver thrust areas include advanced vehicle designs and mine countermeasures.

The Marine Corps has special requirements for its vehicles that differ from those for the Army's vehicles. In particular, operations in the sea and in coastal regions call for vehicles that can be propelled in water as well on land. In addition, the need to carry them by helicopter and inside V-22s imposes size and weight constraints. The Marine Corps reconnaissance, surveillance, and targeting vehicle (RST-V) and the Marine air-ground task force (MAGTF) expeditionary family of fighting vehicles (MEFF-V) are being designed and developed to satisfy these needs, while incorporating advanced technologies from the new Army vehicles.

Autonomous control systems for various vehicles have been developed for all the Services. Some variants of these control systems are expected to be incorporated into the new Marine Corps designs. Special requirements for operation in the surf zone will have to be considered by the ONR researchers providing the technology for overcoming threats close to shore.

ONR has an Organic Mine Countermeasures future naval capability (FNC) program intended to address both Navy and Marine Corps mine countermeasure challenges from very shallow water through the beach exit zone. The ONR Code 353 focus, then, would be on mine countermeasures challenges unique to the Marine Corps that are not already being addressed by the organic mine countermeasures FNC or Army programs and on sensitizing the Army to Marine Corps needs.

The extremely important Marine Corps need for rapid and reliable mine detection has motivated the conduct of many programs in the recent past. These programs have had different degrees of success, but it is fair to say that to date, none have shown sufficient promise for solving this high-priority problem. Littoral remote sensing (LRS) and broader all-source means can be used to locate potentially mined areas both on land and in the sea and to cue mine detection systems, but are effective to only a limited degree. Timely and cohesive characterization of the potential threat remains key.

The variety of situations and media involved makes the mine detection problem very difficult to solve. Mines are found in a variety of configurations and deployments. Some are magnetic, some are nonmagnetic (such as plastics), and some may be buried either deliberately or as a result of shifting sand. Still others may be secured with objects in the surf zone, where they can be obscured by turbidity and other mechanisms. On the other hand, wave action in the surf zone might eventually sterilize some of the mines, thereby reducing the threat, provided that no mines had been recently laid.

Mine detection on dry land is another problem that is generally shared by the Army and, in the case of detection at widely distributed locations, the Special Forces. Standoff mines close to clearings could pose threats to V-22 landings.

PROGRAMS REVIEWED

Joint Mine Detection Technology

The Joint Mine Detection Technology (JMDT) exploratory development program seeks to develop technology for the high-priority capability of remote detection of minefields through the use of imaging sensors. Planned near-term efforts include the development of a multispectral sensor, a laser illuminator, and algorithms for the declaration and characterization of minefields. Short-term transition targets include the UAV-mounted coastal battlefield reconnaissance and analysis (COBRA) system that attempts to detect minefields through image analysis and the Organic Mine Countermeasures FNC.

Findings

The elusiveness of a science-based approach in the myriad of earlier unsuccessful programs to some extent bears witness to the many difficulties associated with this long-standing problem. It would also suggest that choosing the best (or the only) concept available at a given time and then engineering a package for experimentation has not been a successful approach. This failure is very likely due to the lack of continuity of basic research focused on the problem, which requires breakthrough solutions.

The relationship between this program and Army and Navy countermine exploratory development programs was not made clear, although the committee is generally aware of the limited success of the COBRA and the similar Airborne Stand-off Mine Detection System (ASTAMIDS) program and of ONR Code 32's efforts over the last year to cast a wider net for ideas and performers. Apparently, the scope of the JMDT program includes land and sea mines, buried, protruding above the bottom, and on the surface.

The JMDT program, as described to and understood by the committee, did not entail the systematic use of scientific methods expected of an ONR exploratory development program. There seemed little interest in understanding the phenomenon of detection. No data on the statistics of targets and clutter at various wavelengths were presented. The addition of a new wavelength for investigation was justified on the basis that it had not been tried before. Possible differences in performance under sunlight, diffused daylight, and laser illumination were not discussed. The failure to provide a stable test minefield severely limits the value of any particular test result in an apparently haphazard search.

Recommendations

The committee recommends that Code 32 be enlisted to ensure that the goals and approach of the JMDT program, if it continues, are coordinated with those of the Navy and Army programs. All three Services should cooperate in building and maintaining a stable test minefield to which various sensors can be brought.

Given that the intent of this program is to provide broad-area information on the likely locations of minefields in enemy territory, sensor requirements need to consider area coverage rates, false contact rates, and platform survivability in the flight profile required for optimized sensor performance.

ONR Code 353 should leverage other ONR efforts such as exploitation of satellite imagery in the LRS program and assess the degree to which LRS products can be used to cue JMDT-type sensor approaches.

ONR should seek advice on and assistance with instrumentation from other remote sensing organizations. Powerful airborne multispectral and hyperspectral sensors exist—for example, the Hyperspectral Digital Imagery Collection Experiment—and flights of these sensors over a controlled test minefield should produce a database that can be widely disseminated to multiple investigators, who can subsequently determine how much spatial and spectral resolution is needed for minefield detection. Development of prototypes for Marine Corps use can resume once this fundamental knowledge is gained.

The database can also be used by multiple investigators for refining detection algorithms to support timely information dissemination. The Navy Tactical Exploitation of National Capabilities (TENCAP) Office, which has demonstrated powerful multispectral and hyperspectral algorithms for detecting anomalies in clutter, should be consulted in this regard.

The need for a coordinated 6.1 program in the phenomenology for the remote detection of land mines is cited in Chapter 7 in "Recommendations for New Programs" (page 62). The committee recommends that a continuously funded 6.1 program be established. The program would include the categorization of all threat conditions and backgrounds for all mine deployments, e.g., shallow water, surf zone, soft beach, vegetated and barren dry land, and wetlands. In particular, the actual threat of mines deployed in the surf zone over time should be quantified and validated. This involves fusion of a priori surveillance with intelligence information that tracks mines to deployment. As part of the basic research, a comprehensive search over many disciplines should be conducted. In recognition of the difficulty this problem poses, the convening of a blue-ribbon review committee with participants having a broad spectrum of scientific expertise should be considered. Such a committee would be charged with identifying and recommending the most promising candidates for demonstrator-level experimentation and data fusion in all the zones of interest to the Marine Corps.

Reconnaissance, Surveillance, and Targeting Vehicle

The RST-V is being designed and developed as a replacement for the high-mobility, multipurpose wheeled vehicle (HMMWV). The goal is to provide an advanced vehicle that can be carried internally by the V-22 air transport vehicle. It has been designed to have an adjustable hydraulic suspension that can be adjusted from 4 inches for transportation in the V-22 to 24 inches for cross-country operation. To improve fuel consumption and to allow very quiet operation, the vehicle will have a hybrid diesel/electric propulsion system, with electric motors coupled to each wheel and with an advanced storage battery. An automotive diesel engine would be used for long-range operations and to recharge the batteries.

A first technical demonstrator (TD1) is being evaluated and a second one is being fabricated. The TD1 has lead acid batteries, but later versions may have lithium-ion rechargeable batteries.

The ONR managers and their Marine Corps sponsors have specified that the design of the vehicle should provide opportunities for transitioning to several other components of the maneuver and firepower program areas, including the unmanned ground vehicle (UGV), the MEFF-V, and the light armored vehicle (LAV).

Findings

The committee concluded that the technical demonstration of the RST-V is a good 6.3 project and could lead to an excellent replacement, with enhanced capabilities, for the HMMWV, the tri-Service, high-mobility, multipurpose wheeled vehicle that is coming to the end of its service life.

Lithium batteries have very high energy density, and the early primary cells had a reputation for dangerous explosive reactions before, during, and after use. Newer batteries have safety features, such as pressure-relief valves and current-limiting switches, to minimize the danger. The proposed 240-volt rechargeable batteries would have perhaps 80 cells in series, making them particularly susceptible to dangerous reactions. New materials are being developed for the lithium rechargeable batteries, and researchers expect that their safety will be improved, although this has not yet been demonstrated in multicell batteries.

Recommendations

The committee recommends that future consideration be given to other power plants (such as the micro gas turbine) that may have features superior to those of the diesel engine and that the safety features of the lithium battery be carefully evaluated. All new service technologies should also be carefully evaluated with regard to vehicle reliability.

MAGTF Expeditionary Family of Fighting Vehicles

The concept for the MEFF-V utilizes basic chassis, structure, and power/propulsion modules that can be coupled to other components to provide a family of vehicles. The vehicles cover a wide range of configurations, from medical support to tank replacement. The MEFF-V has requirements for swimmability and helicopter transportability, which make it different from the equivalent Army vehicles. The design does, however, utilize technology from Army, Defense Advanced Research Projects Agency (DARPA), and other Marine Corps vehicles (advanced amphibious assault vehicle (AAAV) and RST-V).

Findings

The MEFF-V concept is a very worthwhile project and provides a transition target for 6.1, 6.2, and 6.3 S&T efforts intended to improve vehicle performance.

Recommendations

The committee recommends that the MEFF-V project should be supported. Design assessments should be carried out from time to time to provide assurance that the anticipated benefits of commonality and modularity have been realized.

Autonomous Operations

The Marine Corps seeks autonomous vehicles to increase survivability, reduce casualties, increase combat effectiveness, reduce manpower requirements, and lower logistic burdens. Related efforts include work sponsored by the Office of the Secretary of Defense (OSD) Joint Robotics Program and being conducted by DARPA, the Department of Transportation, academia, and the other Services.

The initial autonomous operations have been planned to have teleremote control to keep humans in the loop while the control algorithms are developed for more difficult conditions, such as the surf zone.

Findings

The application of autonomous systems technology to amphibious assault vehicles (AAVs) and UGVs should provide reconnaissance and surveillance capabilities in dangerous environments and capabilities for decoys and deception. Autonomous vehicles would be particularly useful for urban operations and to clear the surf zone.

Recommendations

The new vehicles (RST-V, AAAV, and MEFF-V) should be designed with control systems that can be converted easily to autonomous operations. The related research and development efforts in ONR, the Department of Defense (DOD), the National Aeronautics and Space Administration (NASA), and the oceanographic community should be monitored.

Simulation-based Acquisition Tools

The benefits of modeling and simulation as a foundation technology applied to a variety of planning tasks and situations are important to the Navy and Marine Corps. They were reviewed by the Naval Studies Board in several recent studies, including *Technology for the United States Navy and Marine Corps 2000–2035*,[1] which reported that modeling and simulation now underlie all aspects of the design and use of military systems and forces.

The ONR Code 353 Simulation-based Acquisition Tools program is intended to enhance the acquisition process for a variety of Marine Corps vehicles. The approach is to leverage existing models of logistics, survivability, and urban operation characteristics. Current plans do not include the integration of cost modeling.

Findings

The Marine Corps, like all the Services, could certainly benefit from a comprehensive simulation-based acquisition tool. However, for the tool to be truly useful in an environment of shrinking budgets, inclusion of cost modeling should be seriously considered. The committee noted the difficulty of integrating disparate models and simulations that were independently conceived and implemented. It also questioned whether the proposed effort was properly classified as exploratory development, and it saw no focus on Marine-unique needs.

Recommendations

The committee recommends that this program be redirected to include only those elements that are unique to the Marine Corps and that it be incorporated in a more generalized simulation-based acquisition tool program that would be of use to all the Services. It also strongly recommends that cost models be included to ensure that studies of trade-offs and the decision-making process include all the necessary elements. The committee further recommends that the project's classification as a 6.2 program be reexamined.

Summary of Recommendations for Maneuver

A summary of recommendations for maneuver is given in Table 2.1.

RECOMMENDATIONS FOR NEW PROGRAMS

The committee has no recommendations for new 6.2 and 6.3 programs in the maneuver area; it recommends in Chapter 7 in "Recommendations for New Programs," (pages 62-63) that ONR 353 set up a sustained, coordinated 6.1 program to establish a scientific basis for the detection of mines in the littorals, particularly remote detection, and for mine countermeasures.

---

[1] Naval Studies Board, National Research Council. 1997. *Technology for the United States Navy and Marine Corps, 2000-2035: Becoming a 21st-Century Force,* 9 volumes, National Academy Press, Washington, D.C.

TABLE 2.1 Summary of Recommendations for Maneuver

Project	Recommendation
JMDT	Enlist other institutions in conducting controlled sensor experiments and in gathering, analyzing, and publishing databases.
RST-V	Consider other power plants such as the micro gas turbine, and evaluate lithium battery safety.
MEFF-V	Support, but periodically reevaluate payoff.
Autonomous Operations	Design new vehicles to accommodate autonomy. Monitor the work of NASA and others.
Simulation-based Acquisition	Ensure Marine Corps specificity; include cost. Reexamine funding category.

CONCLUDING REMARKS

The committee considered most of the programs presented to be of value in meeting future Marine Corps needs. However, this chapter's overview (pages 14-15) identifies a problem that is not being addressed: the transformation of information into knowledge that would allow the Marine Corps to achieve informed maneuver. Information challenges in the urban environment are particularly difficult and, from the briefings provided to the committee, are not being addressed.

The committee identified the need for a coordinated, continuing 6.1 program in the basic science of remote detection of land mines that includes the categorization of all threat conditions. As part of the basic research management, a comprehensive search over a wide area of disciplines should be conducted; consideration should be given to a blue-ribbon committee whose members have a broad spectrum of scientific expertise. Such a committee would be charged with identifying and recommending the most promising candidates for demonstrator-level experimentation in all the areas of interest to the Marine Corps.

3

Firepower

OVERVIEW

Any appraisal of the Office of Naval Research (ONR) Code 353's programs in the areas of firepower or fire support must be made in the context of Marine Corps tactical doctrine (Operational Maneuver From the Sea (OMFTS)), Marine Corps acquisition programs, and ONR future naval capabilities (FNCs) efforts.

Although it is not appropriate to provide a full review of all of the implications of OMFTS in this chapter, several aspects of the OMFTS doctrine are stressed here. Marines are envisaged as operating in a dispersed mode and being capable of providing precise targeting information to sea-based forces so that targets may be engaged by remote platforms. The accuracy of target location should be the criterion against which ONR Code 353's fire support programs are evaluated.

Although they will not mass in heavy concentrations of manpower, individual units will need to be able to defend themselves effectively if engaged by enemy ground forces. The dispersed mode of operation envisaged in OMFTS places a particular strain on the logistic resupply process. Weapon efficiency becomes an extremely important consideration. The more lethal a weapon system is, the fewer its requirements for ammunition resupply. Enhanced weapon lightness and lethality should be the criteria against which ONR Code 353 firepower programs should be evaluated. Camouflage and concealment should be included in this evaluation.

Although the OMFTS doctrine stresses the avoidance of large concentrations of supplies and personnel during the course of extended engagements, naval forces ashore will of necessity establish enclaves that will be targets for attack by hostile cruise missiles and aircraft. Naval forces ashore must have an effective, light, and easily transportable capability for air defense in regions that are beyond the line of sight of the sea-based forces. ONR Code 353's efforts in air defense should be evaluated in this context.

The Marine Corps has a number of acquisition (Milestone III or engineering and manufacturing development (EMD) phase) programs under way in the areas of firepower and fire support. These include the following:

- Target location, designation, and hand-off system;
- The lightweight 155-mm howitzer;
- Javelin;
- Predator;
- Antipersonnel obstacle breaching system; and
- Thermal weapons sight.

ONR Code 353's firepower and fire support programs must also be viewed in the context of Marine Corps acquisition programs of record in these areas. Although these programs have moved beyond the S&T stage of the development and acquisition process, it is certainly appropriate for Code 353 to support enhancements to these systems that can be retrofitted into them so that performance can ultimately be improved.

The FNC endeavor described in Chapter 1 (in "ONR S&T Management," pages 9-10) appears to represent a significant change in the development process. Within ONR, FNC offices will focus resources to ensure that specific desirable future capabilities are developed without diluting this focus by pursuing the discovery programs that will provide longer-term capabilities.

ONR has an FNC program in the area of missile defense. This program appears to be focused on the provision of missile defense for areas that are beyond the line of sight of the Navy's Aegis radar. This effort will support an extended application of the Navy's cooperative engagement capability, the development of a fully active version of the Navy's SM-2 missile, and the integration of these systems with Marine Corps air defense systems. The scope of this effort is clearly beyond the capabilities and responsibilities of ONR Code 353.

PROGRAMS REVIEWED

Enhanced Target Acquisition and Location Program

ONR Code 353's Enhanced Target Acquisition and Location (ETAL) program is dedicated to a significant enhancement to the performance of the Marine Corps target location, designation, and hand-off system (TLDHS) that will permit a forward observer or forward air controller to send digital target coordinates to platforms providing supporting fire. The capability that will be provided by the TLDHS clearly will be central to the execution of the OMFTS doctrine. Unfortunately the version of the TLDHS that is being acquired employs a magnetic compass for the determination of a target's bearing. Aside from the difficulties of maintaining the calibration of such a sensor, it is inherently a sensor of limited precision. Thus, although the TLDHS uses a laser range finder to determine a target's range from an observer whose location is established with the global positioning system (GPS), the TLDHS cross-range precision is poor.

Findings

The ETAL effort is exploring the use of GPS interferometers and miniaturized gyrocompasses so that an azimuth accuracy of 7 miles or less may be achieved against a static target by a man-portable targeting system in favorable visibility conditions. Efforts to date in support of the ETAL objective appear to be proceeding successfully.

The committee is concerned that no S&T funding is projected for man-portable targeting systems beyond FY03. Since the development of an effective man-portable targeting system is central to the

OMFTS concept, the committee suggests that targeting technology should become a core component of ONR Code 353's future efforts. An ETAL-equipped TLDHS will have a limited capability against rapidly moving targets. Furthermore, the time of flight of even moderate-range weapons such as the extended-range guided munition (ERGM) will be between 6 and 9 minutes. During intervals of this length, a mobile target can move far outside of the lethal footprint of a weapon such as ERGM, which currently does not have a data link to provide an update on target location.

Recommendations

The committee recommends that ONR Code 353 should not consider the projected successful completion of its ETAL effort as the last effort in targeting technology in its S&T program. It is strongly suggested that a program to support the targeting of multiple mobile targets be initiated as soon as the budget will permit. Emphasis should be given to the development of a realistic targeting system that will support the use of ERGM and the advanced land-attack missile (ALAM) against multiple mobile targets.

The Objective Crew-served Weapon Program

The Objective Crew-served Weapon (OCSW) program is an Army-led advanced concept technology demonstration (ACTD) that has enjoyed a nominal level of Marine Corps participation. The goal of this program is to produce a lightweight, crew-served infantry weapon that can, in principle, replace the caliber-50 M2 heavy machine gun, the 40-mm MK-19 general-purpose machine gun, and the 7.62-mm M240 machine guns. Because of its greatly enhanced lethality per round, the logistical support for this weapon will be reduced significantly.

The weapon measures the velocity of each round as it proceeds down the barrel of the weapon and uses a laser range finder to determine the range to the target. The fuse on each round is set so that the projectile will detonate at the correct range. Detonation provides a controlled pattern of downward fragmentation that is lethal to anyone in defilade below the burst point of the round. The lethal range of the weapon is about 2,000 meters. In addition, the weapon has a high potential to damage light and lightly armored vehicles at ranges beyond 1,000 meters.

Findings

The development of the OCSW appears to be proceeding successfully, and there is a high probability that the program will satisfy its exit criteria. By the end of FY02, the ACTD phase of the program will be completed. An EMD phase has been scheduled for FY03 to FY07. Production and acquisition are scheduled for the years between FY08 and FY14. To date, the Marine Corps has not decided what actions it will take with regard to acquisition of the OCSW should the current ACTD transition through EMD to production. Some committee members with extensive active-duty Marine Corps experience believe that the Marines will retain the M240 and will use the OCSW as a partial or full replacement for the M2 and the MK-19 machine guns.

From FY00 through FY02, ONR Code 353's contribution to this program is scheduled to be at the $500,000 level. However, the Marine Corps S&T program evaluation group (PEG) has structured a program objective memorandum (POM) for the six fiscal years FY02 to FY07, which shows funding for OCSW rising to $1.8 million in FY02 and, after some fluctuations, stabilizing at $2 million between FY05 and FY07.

Recommendations

Based on the Army's projection for this program, the purpose of the proposed funding contained in the S&T PEG POM 02 is unclear. The committee assumes that the POM 02 funding line for OCSW is a placeholder for an unspecified new start. If this assumption is correct, the committee applauds the decision.

Complementary Low-altitude Weapon System

The Complementary Low-altitude Weapon System (CLAWS) program is conceived of as a Marine expeditionary unit deployable, highly mobile, high-firepower, medium-range surface-to-air weapons system. It is intended to complement the existing Marine Corps short-range air defense (SHORAD) system, which is based on the use of the relatively short range Stinger missile. The concept is to mount six advanced medium-range air-to-air missiles (AMRAAMs) on a high-mobility, multipurpose wheeled vehicle (HMMWV). The complete system will also require the development of an HMMWV-mounted multi-role radar system (MRRS) to provide target detection and an HMMWV-mounted command and control (C2) module.

A milestone 0 decision for CLAWS was made in 1999. Currently, the program is proceeding with concept exploration activities. Although for the purposes of this review CLAWS was listed as an ONR Code 353 program, it is scheduled to be included under the umbrella of ONR's Missile Defense FNC.

With the stand-down of the Marine Corps Hawk capability, until CLAWS becomes operational, Marine forces ashore will have no active air or cruise missile defense except for that provided by the Stinger SHORAD. Although the concept of OMFTS deemphasizes large, local concentrations of Marine resources and manpower, the committee believes that substantial lodgments will be established ashore and that local-area air defenses will have to be established.

Findings

The committee views the problem of mounting AMRAAMs on an HMMWV as a rather straightforward engineering problem that should present few risks. The stressing component of the CLAWS program will be the development of the MRRS. This HMMWV-mounted radar must be capable of detecting and tracking low-observable (e.g., -30 dB), low-trajectory cruise missiles at ranges of 40 miles or more against a background of land clutter.

Although radar systems with comparable performance capabilities have been fielded, their development histories have been long, difficult, and expensive. The committee does not view the management of such a radar system development program as an appropriate activity for ONR Code 353. The committee concurs that the transfer of this development to an ONR FNC office or to a systems command (SYSCOM) office would be appropriate.

Recommendations

In addition to the development of the MRRS, CLAWS requires the development of an HMMWV-mounted C2 module. This C2 module will be required to track targets detected by the MRRS acting as a node in the cooperative engagement capability network. In addition to accepting cues and tracks from remote sensors, the C2 module will be required to report detections and tracks to the information system that maintains the single integrated air picture in the combat area. The committee believes that the

TABLE 3.1 Summary of Recommendations for Firepower

Project	Recommendation
ETAL	Look beyond ETAL to targeting multiple mobile targets.
OCSW	Maintain a funding line for new starts.
CLAWS	Transfer to FNC.

development and packaging of such capabilities are technically feasible. However, it also believes that the development of this module will be a sufficiently complex engineering problem that it, too, should be transferred from ONR Code 353's cognizance to either an ONR FNC office or a SYSCOM office.

Summary of Recommendations for Firepower

A summary of recommendations for firepower is given in Table 3.1.

RECOMMENDATIONS FOR NEW PROGRAMS

Small Arms

The committee is aware that the Army has been assigned the lead in the development of small arms and infantry weapons. Although the OCSW appears to be a spectacular and successful weapon development, the committee is concerned that other basic infantry weapons such as hand grenades and mortars have shown little if any advance since World War II or before. Although ONR Code 353 cannot institute new developments in these areas, it can support studies of and concept formulations for other advanced infantry weapons. The committee hopes that ONR Code 353 would then be able to persuade the Army to support future joint developments of other advanced infantry weapons.

The Marine Corps has been assigned lead service status for nonlethal weapons. The committee was struck by the fact that ONR Code 353's program does not have a component assigned to supporting work in this area. The committee believes that nonlethal weapons will have an important role in many of the future situations in which Marines may become involved. The difficulty with nonlethal weapons is that situations for which they are appropriate may develop within moments into situations where lethal weapons are more appropriate (requiring, for example, a transition from the use of rubber bullets to real bullets, and so on). ONR Code 353 should take the lead in developing concepts for new weapons of controllable lethality that will allow a rapid transition from nonlethal to lethal modes of operation.

Overall, the committee was impressed with the progress that has been made in the OCSW ACTD. As this ACTD's end approaches, the committee recommends that ONR Code 353 should structure a POM 02 program that concentrates on the following:

- R&D related to concepts for weapons of controllable lethality;
- The development of new concepts for conventional infantry weapons such as mortars, hand grenades, and rifle grenades; and
- Countersniper weapons.

> **Box 3.1**
> **Summary of Recommendations for New Investments in Firepower**
>
> - Concepts for weapons of controllable lethality
> - New concepts for mortars, hand grenades, and rifle grenades
> - Countersniper weapons
> - Sensor systems to support counterbattery weapons

All such proposed developments should concentrate on weapons that have reduced logistic support requirements.

Counterbattery Systems

The committee was somewhat surprised that the program of ONR Code 353 does not support technology that will lead to the development of efficient counterbattery capabilities. Although some might argue that ERGM and OCSW are in some sense inherently counterbattery weapons, they are not supported by a targeting system that can locate—based on the ability to backtrack the trajectory of hostile rounds to their launch point—a battery in defilade or one that is beyond the line of sight.

Although past efforts to develop such a capability have been uniformly unsuccessful, the committee does not believe that the problem is inherently intractable. The committee would encourage ONR Code 353 to explore new approaches to this difficult but tactically important problem.

Summary of Recommendations for New Investments in Firepower

A summary of recommendations for new investments in firepower is given in Box 3.1.

CONCLUDING REMARKS

The committee found ONR Code 353's efforts in the areas of firepower and fire support to be well managed, productive, and appropriate. All of the programs in these areas are likely to move beyond Code 353's area of responsibility within 1 to 3 years. The lack of substantive discussion of future efforts that will replace maturing current efforts concerned the committee, however.

Firepower and fire support are clearly areas of considerable importance to the Marine Corps. The committee believes that ONR Code 353's program should be the incubator for new ideas and concepts that exploit advances in technology. In the discussion above, the committee suggests that Code 353 might explore new concepts for weapons of controlled lethality. That suggestion hardly exhausts the novel weapon concepts that might be explored. Advanced technology might be applied to almost every infantry weapon. As an example, anyone who has ever served in an infantry heavy-weapons company will agree that weapons are too heavy. Might not advances in materials science be exploited to reduce the weight of the base plate of an 80-mm mortar? Although such efforts would not yield a Nobel Prize, they would earn the gratitude of the Marine infantrymen who lug such devices around a battlefield.

As another example, the committee suspects that the range, accuracy, and lethality of 60- and 80-mm mortars, along with the number of rounds needed to zero in on a target, could be improved by applying modern technology to systems that are not intrinsically different from their World War I antecedents. Although not insisting on specific fire power and fire support improvements, the committee recommends most strongly that Code 353 should begin to structure replacement programs for the current successful programs that will soon be removed from Code 353's portfolio.

4

Logistics

OVERVIEW

The U.S. Marine Corps is moving toward the Operational Maneuver From the Sea (OMFTS) concept, which relies heavily on sea-based logistics with a minimal logistics footprint ashore. This sea-based logistics system is a time-critical, transportation-intensive system. Requests for supplies from the warfighters ashore will be quickly assembled into shipments aboard ship and transported to the beach and beyond by airlift (V-22 and helicopters) and/or surface lift (advanced amphibious assault vehicle (AAAV) and the landing craft, air-cushioned (LCAC)). This logistics concept was reviewed by the Naval Studies Board in 1998, and several significant findings were documented in the report *Naval Expeditionary Logistics: Enabling Operational Maneuver From the Sea.*[1] One significant finding was that the OMFTS concept and key parameters such as the desired level of combat capability and the operating distance ashore were not yet fully defined. As a result of these uncertainties, it was not possible for that study to determine the critical shortfalls in the logistics system and where improvements were most needed. Nonetheless, the committee observes that Code 353 research should focus on improving the Marines' ability to provide logistics support from sea bases.

For even a fairly light Marine landing force ashore, comprising 6,800 Marines with current equipment, the resupply requirement was 490 tons/day. That study found such a supply rate could be maintained for a ship-to-warfighter distance of 125 miles, but not much farther with currently planned lift capabilities. Knowing about the distribution of supplies required by that landing force, as shown in Table 4.1, is necessary for understanding some of the findings of the current study committee. Fuel is clearly the major logistics driver, consuming about 46 percent of the lift required, but water is a close second, followed by munitions.

Based on requirements like these, the committee recognizes the high value of programs that can significantly reduce the OMFTS logistics requirements. Specific programs include those for the recon-

---

[1]Naval Studies Board, National Research Council. 1999. *Naval Expeditionary Logistics: Enabling Operational Maneuver From the Sea,* National Academy Press, Washington, D.C.

TABLE 4.1 Landing Force Resupply Requirements

Item	Quantity (tons/day)	Percentage of Total
Fuel	225	45.9
Water	190	38.8
Munitions	33.5	6.8
Other	26.5	5.4
Food	15	3.1
Total	490	100

naissance, surveillance, and targeting vehicle, which uses a dual-power propulsion system and offers significant fuel savings, reducing consumption for current reconnaissance vehicles by a factor of 1.7 to 2, and the objective crew-served weapon, which potentially offers a 20-fold reduction in the weight of munitions that have to be transported. The committee also recognizes that logistics is typically in a reactive mode, given the task of transporting and maintaining equipment that has often been designed without logistics constraints as a key design consideration, and it encourages the Marine Corps logistics community to take a more proactive role in the future.

PROGRAMS REVIEWED

Bulk Liquids

The Bulk Liquids program includes transport of both fuel and water, but the emphasis of the projects presented was on fuel and concentrated on methods of facilitating transportation and automatic sensing of fuel quantity and fuel quality. One project is developing an expeditionary fuel system (EFS) consisting of ten 400-gallon tanks on a lightweight pallet system that can be easily reconfigured in a variety of ways. Another project is exploring the use of commercial off-the-shelf sensors and communications technologies to automatically monitor and report on the quantity of fuel available in the field. The third major project is exploring a variety of sensing techniques for hand-held instruments to determine the quality of fuel in the field.

Findings

The committee saw the EFS more as an engineering effort than a science and technology issue. However, the system does reduce the deadweight that has to be transported and the storage space required for the containers aboard the ship, and it adds considerable flexibility to field operations by allowing partially empty fuel containers to be consolidated. The program for implementing fuel-quantity sensors appeared to be an adaptation of commercial technology but was seen as important in developing an automated logistics picture of the battlefield. However, there were concerns about interference and interoperability of the fuel sensor radio-frequency communication links with other communications activities as well as the potential vulnerability of the EFS to information warfare by the adversary. The fuel-quality sensors were seen as an important area of investigation: the use of host nation and captured fuel supplies can significantly reduce logistics transportation demands, but fuel quality is a critical issue in such use. (See, for example, in the Chapter 7 recommendation on the development of fuel cells (page 61) a comment on the need to measure sulfur contamination in fuels.)

Recommendations

The committee suggests that the EFS work be completed and transitioned. The fuel-quantity sensors and communications systems should be examined for their compatibility with other field communications systems as well as for their security. The exploration of hand-held sensor technologies for determining fuel quality is considered an important effort and should be continued. Additionally, the Bulk Liquids program should devote more attention to water, which places the second largest demand on the logistics transportation system and will have to be transported largely from ships to support OMFTS forces.

Corrosion Technologies

The aim of the Corrosion Technologies program area is to demonstrate new corrosion prevention technologies or corrosion-resistant materials to be used by designers of future Marine Corps equipment. The approach is three-pronged:

- Conduct exposure tests of materials and coatings in the marine environment,
- Evaluate accelerated test procedures for U.S. Marine Corps applicability, and
- Coordinate with other corrosion protection programs in the commercial and military worlds.

Findings

The committee recognizes that corrosion is a significant cost issue for the Marine Corps and that it is a difficult problem to address in up-front designs that are typically bought jointly with another Service such as the Army, which does not face the same corrosion issues. From this perspective, ONR's Corrosion Technologies program is seen as taking a proactive view in trying to help reduce future corrosion. However, the committee is concerned that the approach may be repeating much of the work that has been done previously by industry, that it is not taking full advantage of the available scientific knowledge of the processes and physics of corrosion, and that it is spread too thinly to produce a lasting effect.

Recommendations

The committee recommends that the Corrosion Technologies program concentrate on a specific system, preferably the AAAV. Identifying specific AAAV corrosion problems from discussions with operators and specific technical approaches based on extensive interactions with the corrosion scientific community should become the first priority. The committee recommends that the results of those efforts be used in planning for and testing specific, high-payoff fixes for some of the more expensive and serious AAAV corrosion problems.

Modeling and Simulation

A significant modeling effort for OMFTS logistics has been under way for some time in the development of the tactical logistics distribution system (TLoaDS). This effort is a discrete time-simulation model that uses logistics algorithms to conduct deterministic or stochastic simulations of OMFTS logistics systems. The proposed work would continue the development and use of this modeling tool and extend modeling to include replenishment of the ships at sea.

Findings

The Naval Studies Board report *Naval Expeditionary Logistics* strongly recommended the use of modeling and simulation tools. This committee concurs with that recommendation and emphasizes that the modeling efforts should be used to address important operational and design issues. The use of models like TLoaDS and their expansion to address additional facets of the OMFTS logistics problem should be continued.

Recommendations

The committee recommends that the set-up time for TLoaDS (currently 2 to 3 months for the front end) be significantly shortened and simplified to make the tool easier to use. These models should be used to assess the effects of vulnerability of the logistics airlift to enemy action in terms of delivery shortfalls and time delays in delivering supplies to the warfighters inland, as well as the impact of bad weather on the at-sea replenishment of the naval force. Vulnerabilities and requirements for aircraft survivability/countermeasures should be communicated to the appropriate development programs.

Small Unit Logistics

The Small Unit Logistics (SUL) advanced concept technology demonstration (ACTD), started in 1999, is concentrating on supporting logistics command and control, plan execution, and logistics visibility for the small unit at the tactical level. The emphasis of the ACTD is on providing the commander a common tactical/logistics picture at the mission execution phase. The SUL ACTD goals are a proof-of-concept demonstration of improved logistics command and coordination and a smaller footprint for tactical unit logistics. The idea is to use client server architecture to link a number of different legacy databases with tagging devices to allow locating equipment and supplies in real time and to make this common tactical picture available via a communications network based on existing systems such as the single-channel ground and airborne radio system (SINCGARS) radio.

Findings

The committee is concerned about the limited bandwidth available with current communications systems and the amount of data that has to be transmitted in the SUL concept. The decision to send only changes in database contents in order to keep bandwidth demands at acceptable levels has potentially serious consequences. Errors in receiving the database updates or failure to obtain the updates at all can lead to a slow degradation in database consistency, ultimately producing a confusing, incompatible battlefield picture that can result in incorrect decisions. The SUL plan to integrate a variety of legacy databases is also disturbing, since it raises a number of database incompatibility problems and tends to perpetuate a database interoperability problem. Moreover, the committee notes that many critical decisions about how to best deliver logistics support must be made in very short periods of time during OMFTS operations and believes that decision aids may provide significant support to the operators under these conditions.

Recommendations

The committee suggests that the SUL ACTD consider the use of advanced error-correcting codes for ensuring the accuracy of database updates and that it develop some methodology to ensure that missed updates are recognized and rerequested by all affected parties. It also recommends that the program, rather than simply integrating existing legacy databases, consider selecting a next-generation database system to be used for one of the legacy databases as a demonstration of the power of new database management systems. Such a step could help solve interoperability problems if the most difficult legacy system were selected for conversion and also used as a transition platform to encourage updating the other legacy database systems by the community. Although the committee recognizes that the current communications structure must be used for some time, proof-of-principle demonstrations of the value of higher-bandwidth communications systems would be a valuable adjunct to the ACTD. The use of the ultra-wideband radios, discussed in the 6.1 research area, should also be considered. The committee also recommends increased attention to the development and testing of decision aids to ensure the effective use of the data by Marine Corps logistics commanders.

Summary of Recommendations for Logistics

A summary of recommendations for logistics is given in Table 4.2.

TABLE 4.2 Summary of Recommendations for Logistics

Project	Recommendation
Bulk Liquids	Complete and transition EFS.
	Continue sensor exploration.
	Consider water.
Corrosion	Focus on AAAV.
Modeling and Simulation	Simplify TLoaDS setup.
	Assess airlift vulnerability.
SUL	Use error-correcting codes.
	Devise means of synchronizing databases.
	Develop and test decision aids.

RECOMMENDATIONS FOR NEW PROGRAMS

Vulnerability of Sea-based Logistics Train

The committee is concerned about two potential vulnerabilities of sea-based logistics. In the airlift portion over land occupied by the adversary, the logistics aircraft may be particularly vulnerable to antiaircraft fire, especially from man-portable systems. On the sea-based side, bad weather may degrade or even halt the ability to assemble shipments aboard ship and transport them by air. Although the Marine Corps does not fund aircraft development and aircraft countermeasures, it may have requirements for aircraft survivability/countermeasures that are not being addressed by the appropriate systems development programs. The committee is concerned that unless the Marine Corps takes the initiative in this direction, there is a high likelihood that no one else will address its requirements. Accordingly, the committee recommends that modeling analyses using programs like TLoaDS be used to assess the impact of aircraft vulnerability to enemy action. The resulting Marine Corps requirements for aircraft survivability/countermeasures should then be communicated to the responsible programs. Similarly, TLoaDS or similar models should be developed and used to assess the impact of bad weather on the ability to unload shipboard containers, form logistics delivery packages on board in a limited space, and load those packages onto aircraft for shipment to the warfighters ashore.

Integrated Approach to OMFTS Logistics

OMFTS requires a demand-based logistics system that integrates many different logistics elements into a robust and flexible system. These different logistics elements include transportation of supplies to the theater of operations from the continental United States and other major supply centers around the world, transfer to logistics support ships in theater, repackaging into loads required by the warfighters, and transport forward and delivery to the warfighters. This is a complex system problem, with multiple opportunities for trading off requirements and capabilities among different system elements, as well as among warfighting elements. Moreover, developing a logistics system that reliably supports the OMFTS concept requires that the scope of OMFTS itself be more clearly delineated in terms of force levels, distance inland, and mission duration.

An overall systems perspective needs to be taken toward the OMFTS logistics system. Although the current efforts reviewed by this committee do contribute to an OMFTS logistics system, it is not clear that they are the most cost-effective or the most critical. This comprehensive systems perspective should include interfaces with the commercial logistics systems, the transportation and transfer of equipment and supplies to the OMFTS logistics ships, equipment packaging, flexible repackaging, and transport forward to the forces inland. Such an assessment should also consider the appropriate OMFTS logistics platforms and particularly the requirements for a sea-based logistics platform to support OMFTS. The assessment should also address the expected airlift requirements from the offshore ships to the inland forces, including the number of airlift platforms, their vulnerability, and needed redundancy.[2] Finally, the assessment should identify areas with the potential for high-value trade-offs, where

---

[2]The present number of aircraft cannot support an active tactical scheme and the logistical requirements of a larger (Bn+) force at the ranges envisioned in OMFTS. The movement to inland sites must be addressed as a logistical problem and reviewed in the same way as the landing craft/causeway and the AAAV in traditional surface movement.

> **Box 4.1**
> **Summary of Recommendations for New Investments in Logistics**
>
> - Vulnerability of sea-based logistics train
> - Shipment development/item location in shipboard environment
> - Innovative approaches to water usage

reductions in fuel, water, and munitions requirements can significantly reduce demands on the logistics system. These reductions in demand may come from a variety of factors such as more efficient vehicles, more accurate targeting, and more accurate weapons delivery. This type of overall systems assessment would help the logistics community to play a much more proactive role in the future.

Innovative Approaches to Water Usage

Water accounts for the second largest demand made on logistics transportation for OMFTS. The Naval Studies Board report *Naval Expeditionary Logistics* indicated that per-person usage in a Marine Corps landing force was 7 gallons per day. Discussions during the review also indicated that all the water consumed by an OMFTS landing force would have to be transported from the ships. In part, this transport requirement was driven by the need to keep up with fast-moving troops. Even though techniques exist for purifying the water from indigenous sources such as streams and lakes, the production rate may not be high enough to keep pace with the warfighters. In other environments such as the desert, sources of water are very limited, and the warfighters must rely on having it transported. As a consequence, the committee recommends a two-pronged approach. First, a program to quantify options to reduce nonpersonal consumption of water by recycling and reuse or alternative methods should be considered. Even a 10 to 20 percent reduction in the amount of water required would significantly reduce transportation demands. Second, efforts to improve the throughput of water purification techniques and reduce their weight and power requirements should be pursued, leading to new systems that can convert water from indigenous sources into useful supplies for the warfighters in an OMFTS scenario.

Summary of Recommendations for New Investments in Logistics

A summary of recommendations for new investments in logistics is given in Box 4.1.

CONCLUDING REMARKS

Logistics is critical to the success of OMFTS and should be given more attention within the ONR research programs. The Expeditionary Logistics future naval capability was not briefed to the committee, and it may address some of the issues identified above as gaps. However, it is clear that many logistics-critical elements must be addressed for successful sea-based logistics, and closer coordination with the design programs and the other programs in this S&T area is needed to help reduce the logistics demands as part of the up-front design.

5

Training and Education

OVERVIEW

During the 1990s, the National Research Council conducted three studies on the training of Service personnel. The studies' findings and recommendations, reproduced in Appendix B, provide relevant background for this review of the Office of Naval Research (ONR) Code 353's training and education science and technology (S&T) endeavors.[1]

The four main thrusts in this technological imperative are virtual environment training, live simulation for Military Operations in Urban Terrain (MOUT), Marine air-ground task force (MAGTF) federated object model (FOM), and tactical decision games. More than 50 percent of the financing in training and education is being spent on the first thrust—virtual technologies and environment (VIRTE).

VIRTE's aim is to provide realistic combat simulators for a variety of missions and to complement live training in order to enhance the performance of individual warfighters and small teams. The goal is to make training more flexible, less costly, and more efficient. In contrast with flight simulations and combat and weapon systems simulations, simulations envisioned in this thrust employ immature technology that is steadily improving.

The relationship of VIRTE to the Capable Manpower future naval capability (FNC) is unclear, as is whether ONR Code 353 will have a major role in its management.

The live training for MOUT will provide capabilities for instrumented combatants, urban terrain, and infantry weapons. Its goal is to enhance force-on-force training with high realism for improved performance.

---

[1]Table B.2 in Appendix B shows the technology-based training capabilities expected to be available in 2035. The capabilities sought are evolutionary, not revolutionary. They reflect the eight areas of development likely to change the nature of Navy and Marine Corps training and are outlined along with more than a dozen key enabling technologies. The current, relatively small Marine Corps training and education program is limited largely to one of the eight areas of development—virtual reality—and employs only some of the enabling technologies.

The objective of the MAGTF FOM thrust is to develop a prototype FOM standard for simulator integration to support collective training. Its goal is to deliver a consistent view of the battlefield and sensor interactions for each participant.

The long-term objective of the tactical decision games is to provide network-compatible PC war games that are realistic and engaging. This thrust employs proven, commercially available technology and exploits the wide use of war games currently sold.

PROGRAMS REVIEWED

Small Unit Tactical Training

The stated objective of the Small Unit Tactical Training (SUTT) program effort is "to develop and evaluate virtual environment technologies supporting a training concept to permit individual, team, and interteam distributed training in a synthetic environment." The requirement for this program is to better prepare warriors to meet the physical, emotional, and intellectual demands of modern battle at a time when pretraining budgets and time for live training are decreasing. The effort seeks to improve the quality of interaction provided by virtual environments (VEs) and to develop a reusable VE that will permit interactions in a coherent battlefield. The SUTT program itself ends in FY02, but three follow-on demonstrations are being planned between FY02 and FY07 that move progressively from platform simulation to close-quarters battle and up to full-spectrum combat.

Findings

The committee believes the SUTT program to be extremely important, well-leveraged by other VIRTE efforts, and unique in its effort to create a fully immersive perceptual illusion for the trainee, incorporating realistic interactions. The committee fully supports this effort but also recognizes that little is being invested in basic research and that the program is technology-limited.

Recommendations

The committee recommends that efforts be made to quantify payoffs against baseline performance and against other alternatives. It recommends that a formal evaluation plan be developed at this time that explores cost-effectiveness goals.

Real-Time Motion Capture for Virtual Reality

The goal of this small 6.1 and 6.2 virtual locomotion research effort is to provide natural, direct control over simulated locomotion for realistic maneuver through VEs.

Findings

The committee was very impressed with the program and its execution to date. The planned comparative evaluation of three virtual locomotion techniques for a number of tasks is a needed step in demonstrating relative values quantitatively. Although this effort appears to be a very desirable enhancement of VIRTE, its continuation seems to be in doubt.

Recommendation

Budgetary constraints aside, this effort should be continued and the cost-effectiveness of higher fidelity should be demonstrated.

Wearable Computers

The objective of this effort is to provide untethered motion and movement for individuals performing in a virtual environment and to provide real-time access to electronic technical manuals. The untethered system will provide a more realistic training environment. The technical approach leverages a commercial off-the-shelf (COTS) optical three-dimensional input and tracking device.

Findings

This appears to be another worthwhile addition to VIRTE, but again there was no mention of an evaluation plan or comparison with a baseline.

Recommendation

The effort should be continued through demonstration and should include evaluations.

Rapid Synthetic Environment Database Generation, Small Business Innovative Research, Phase II

The objective of the Rapid Synthetic Environment Database Generation (RSEDG) program is to develop a prototype of a near-real-time, automated, three-dimensional model-generation capability to process video scenes into three-dimensional images. This technology will be applied to VIRTE trainers.

Findings

The committee believes that if the development is successful, it may help to expand three-dimensional applications and enrich VE training. In the brief presentation, it appeared that the technical approach was to utilize the most promising available technique rather than develop a new technique or advance existing techniques.

Recommendation

The effort should be continued through demonstration and should include evaluations.

K-band Shoot Through Obscuration and Training Scoring System Advanced Technology Demonstration

The objective of this program is to enhance the combat training ability of the operational range instrumentation system (RIS). K-band shoot through obscuration (STO) may overcome the limitation of utilizing the multiple integrated laser engagement system (MILES) in an obscured environment. Currently MILES cannot record a target pairing when impeded by a variety of obscurants.

Findings

The committee notes the current success of MILES at the National Training Center in an unobscured environment and could perhaps be persuaded that despite the general need to ease OMFTS logistics by maintaining a high ratio of hits to fires, there is a need to score fires directed at obscured targets. However, the committee notes that a weapon-mounted K_a-band antenna will encompass a diameter of only a modest number of wavelengths and consequently have a rather coarse beam. Such a system would not score weapon/target pairing accurately except at the shortest ranges.

Recommendation

The committee recommends withholding approval to build hardware until the angular accuracy of the system is calculated and the Marine Corps Combat Development Command (MCCDC) affirms that the predicted performance would be useful.

Military Operations in Urban Terrain

The objective of this research is to enable the tracking of individuals, without losing contact, in a MOUT environment.

Findings

The committee believes that the effort could enhance live training for MOUT through instrumentation. Currently, two competing approaches are being developed, and transitioning to the MOUT environment could occur at the end of Phase II in FY02. No additional research resources were indicated.

Recommendation

This effort should continue through transitioning and evaluation.

MAGTF Federated Object Model

The objective of this effort is to develop a prototype FOM standard for simulator integration. As a required interoperability standard, it will establish standard interface and interoperability requirements for combat system simulators. The key technical challenges noted were identifying a common synthetic environment and developing a FOM prototype capability in an open architecture demonstration. The effort is in an early stage and when completed will be mandated for use by FY03.

Findings

The committee agrees that the lack of interoperability among independently developed simulations is a thorny problem, notes that writing and mandating a standard are not S&T endeavors (although demonstration of its implementability arguably could be), that the provision of a common real-time interface does not ensure correct interoperation, and that the proposed schedule is very ambitious. The committee also notes that ONR recently made a corporate decision to embark on a major multidepartment effort to tackle some of the hard problems in simulation technology, including efforts in Code 31 on techniques for the integration of disparate models and simulations.

Recommendation

The committee recommends that the task force that formulated the ONR corporate modeling and simulation initiative review this program, evaluate the likelihood that it can meet its goals, and recommend any needed modifications to the program plan.

Marine Air-Ground Task Force XXI and Marine 2000 Close-Combat Tactical Decision Games

The objective of this effort is to develop low-cost individual and multiplayer tactical decision games developed from commercial PC games.

Findings

The enabling technologies are available. The committee believes that the adaptation of commercial games is a good idea. The program as presented is needed, but very few new technologies are being developed and it is unclear who is doing the current technical work. Nevertheless, the program appears to be the beginning of a significant means of increasing training effectiveness and efficiency by converting standard forms of training to technology-based distributed training. The adaptation of commercial games for decision making is a challenge.

Recommendation

Evaluation plans that include objective measurement against baseline conditions should be incorporated into the effort.

Summary of Recommendations for Training and Education

A summary of recommendations for training and education is given in Table 5.1.

TABLE 5.1 Summary of Recommendations for Training and Education

Project	Recommendation
SUTT	Quantify payoff.
	Develop a formal evaluation plan.
Motion Capture	Continue, but demonstrate cost-effectiveness of fidelity.
Wearable Computers	Continue through demonstration; include evaluation.
RSEDG (SBIR Phase II)	Continue through demonstration; include evaluation.
STO Scoring	Withhold approval until angular accuracy is verified.
MOUT	Continue through demonstration and evaluation.
MAGTF FOM	Review, reevaluate feasibility, and adjust as indicated.
MAGTF XXI and TDGs	Incorporate evaluation plans.

RECOMMENDATIONS FOR NEW PROGRAMS

The MCCDC identified five imperative capabilities to enable OMFTS. One of these areas is training and education. Other key areas generally included in research on human performance are the human-machine interface (human factors engineering) and selection and assignment. A more complete systems view of human capabilities in Marine Corps operations is needed—one that takes into account the interactions among such major factors as selection, training, and ergonomic design, and how they contribute to the overall competence of individuals and teams. In an expanded research program, it would be desirable to conduct cost-effectiveness analyses to establish priorities among the areas of research competing for scarce budgetary resources.

Four new training applications are recommended for emphasis in future technological developments.

General Military Training

Standardized training packages covering essential military knowledge and skills should be available at any time within a training cycle at every computerized workstation at each Marine's convenience. Cost-effectiveness analyses would be required to determine the optimal balance between residential, mission operational specialty preparation training, and on-the-job training.

Proficiency Training

The skills of operators and maintenance personnel should be greatly enhanced through the use of advanced embedded training capabilities.

Team Training

The capability for extensive networking of simulators and microwave linkages to allow land, sea, and air units to train together at the same time against the same mission objectives should be developed. Study and modeling of cognitive processes important to OMFTS should be undertaken.

Limitations of Close-Combat Computer Games

Some members of the committee believe that U.S. superiority in air-to-air combat, despite rough parity in aircraft and weapon capabilities, results from the heavy U.S. investments in training flight hours, instrumented training ranges with reconstruction capabilities, and VEs with expensive, high-fidelity displays. Other nations cannot or choose not to make these investments.

A possible research topic, either for the training and education program or for the 6.1 program, would be to assess the limitations of training games based on commercial computer technology in order to determine whether the inevitable worldwide dissemination of this technology threatens U.S. combat superiority.

Summary of Recommendations for New Investments in Training and Education

A major part of the challenge in developing military (or consumer) gear is to make it reliable and user friendly. The impression is that developers do a fairly good job when pressed. To do it right

> **Box 5.1 Summary of Recommendations for New Investments in Training and Education**
>
> - General military training
> - Proficiency training
> - Team training
> - Limitations of close-combat computer games
> - A long-range research and technology plan for overall training and education that includes the rationale for selecting programs

requires that the developer and the operator work together. Experience with computer games and simulators has enabled high school graduates to operate at a high level many complex systems with reasonable amounts of additional training. Another important aspect of this issue is the close involvement of the contractor with support for the system: maintenance, training, software support, and so forth. If all these aspects are handled correctly, which is a challenge, then the education levels of available troops are probably sufficient. But it will take work, and, in some aspects, doing things differently. Although the best results will probably come from appropriate designs and tests and use of the right acquisition approach, some S&T efforts probably could help with the process and ONR might be a good place to take a look at the fundamentals. A summary of recommendations for new investments in training and education is given in Box 5.1.

CONCLUDING REMARKS

The main issue being addressed in work on training technology is how to transform conventional forms of training into advanced technologies that can be distributed to the field for training individuals and teams. Such technologies need to be effective and efficient and tailored to Marine Corps needs. The training technology currently being developed and the new topics suggested by the committee capitalize on improvements and applications such as collaborative virtual environments, interactive distance learning, and embedded training.

Research has shown that technology-based training can increase performance much more successfully than centralized training and can reduce training time and training resources.[2] Additionally, technology-based training is accessible to and motivates the student and the warfighter.

The S&T programs reviewed are clearly needed and conceptually sound. It is apparent that the Marine Corps must leverage external efforts because of severe budgetary constraints. Such leveraging is being accomplished through partnerships, cooperative agreements, and keeping current with developments. Perhaps because of the need to leverage and the newness of the efforts, the committee could not identify a coherent, integrated, or consistent program. A long-range research and technology plan that includes the rationale for selecting programs is needed for the overall training and education area.

---

[2]See, for example, Fletcher, J.D. 1990. *The Effectiveness of Interactive Videodisc Instruction in Defense Training and Education*, IDA Paper P-2372, Institute for Defense Analyses, Alexandria, Va.

If VIRTE management transitions to the Capable Manpower FNC, Code 353's responsibility will be to manage a discovery program in training technology. Much of the existing Code 353 program consists of experimenting with available technology, a function that could be assumed by the FNC if not by the Marine Corps Warfighting Laboratory. Of the programs reviewed, only the virtual locomotion research effort seemed to be pushing technological frontiers and to have a coherent research plan that explicitly included evaluation.

Demonstration of new capabilities is a requirement, but training technologies must also be shown to be cost-effective, adaptable to differing individual abilities, motivating to individuals and teams, and applicable to new problems. To accomplish this requires careful attention to the early development of training evaluation plans as an integral phase of the overall developmental process.

6

Command and Control

OVERVIEW

Although its basic requirements for command and control (C2) systems resemble those of other land forces such as the Army, the Marine Corps has several specific needs, driven by its unique missions, that are unlikely to be addressed by other Services. The Operational Maneuver From the Sea (OMFTS) doctrine, for example, places demands on a C2 system that are very different from those imposed by doctrines for larger, land-based forces. Indeed, success in OMFTS depends on meeting unique technical requirements for achieving overall information superiority, not just its C2 components, a larger regime considered in this chapter.

In OMFTS, a relatively small force of Marines will launch an assault from ship to shore, will decide on target landing locations while in transit across the water, and may penetrate deep inland with quite small, highly mobile groups. Such a force's relative deficit in numbers and firepower is, according to the OMFTS concept, more than compensated for by knowledge of the battlespace, by agility, and by excellent and timely command and control that can, for example, bring down sea-based fire on enemy targets upon command from the Marines.

Thus, OMFTS places a premium on timely situational awareness and assured communications, even in the face of jamming. Furthermore it requires that such services will be provided without any of the traditional impediments associated with command, control, communications, computing, and intelligence (C4I) systems such as heavy, slow-to-set-up microwave antenna masts, large satellite ground terminals, and so on. This demand for excellent C4I services without the attendant bulk of traditional systems is not entirely Marine Corps-specific; the Special Forces have this requirement, and the Army, too, is trying to become lighter and faster. However, the Marines Corps carries this concept much farther than the Army and as a result has specific requirements for these systems.

From a technical viewpoint, OMFTS poses the following Marine Corps-specific challenges:

- Marines must be wedded to naval C4I systems, more or less closely, until the moment that they launch for the shore.

- At that moment, Marine craft (e.g., the advanced amphibious assault vehicle (AAAV)) must transition to their own, organic wireless communications for their maneuver to land. Their wireless networks must enable the Marines to remain in close contact with naval, theater, and even national systems that can supply the requisite knowledge to commanders.
- During their drive across the sea toward land, commanders in transit (e.g., in AAAVs) must receive accurate, detailed, up-to-the minute intelligence on enemy dispositions and the current locations of the Marines. They must also be able to give commands to their dispersed forces with high probability of success.
- Those Marines that penetrate deep inland must rely on some kind of overhead relay, such as unmanned aerial vehicles (UAVs) or satellites, in order to maintain communication with Marines ashore and the naval forces afloat, to call for fire, and to continue to receive timely intelligence and situational awareness that permits the informed maneuver called for in Chapter 2's overview (pages 14-15).
- While maneuvering inland, these Marines may need to enhance their information superiority by conducting deception and information operations or by having others conduct such operations on their behalf.
- Finally, once other land forces such as the Army appear on the scene, the Marine units may need to plug their C4I systems into the Army's in preference to relying on the purely naval systems. This will allow them to become an integrated part of the Army's situational awareness picture.

Each of these challenges is unlikely to be adequately addressed by another Service but must be successfully resolved for OMFTS to work as envisioned. Each is therefore a natural priority for Marine Corps R&D in C4I.

In the judgment of the committee, the Marines have done a good job of experimenting with prototype C4I systems in OMFTS-type exercises. The Extending the Littoral Battlespace (ELB) advanced concept technology demonstration (ACTD) is an excellent case in point. Here the Marines have made a prototype of the C4I network needed for OMFTS and have employed it in their exercises. The ELB prototype, although by no means an operational system, allows a glimpse of the future world in which new, radio-based networks will provide data communications for dismounted Marines, very agile command centers, and long-haul communications for forces deep inland via airborne relays back to ships afloat.

As discussed in Chapter 8, the committee applauds this experimentation despite the severe limitations of the prototype. The Marine Corps is also urged to take this approach one step further. The committee recommends that ONR Code 353's main focus in C4I should be closely tailored to those Marine Corps-specific requirements imposed by both OMFTS and Military Operations in Urban Terrain (MOUT) scenarios. This general philosophy has guided the committee both in its critiques of the existing Code 353 programs and in suggestions for new directions.

PROGRAMS REVIEWED

Command and Control Components for a Harsh Environment

This program aims to help the Marine Corps procure commercial off-the-shelf computer equipment that will better withstand the harsh environment in which Marines' C2 equipment must operate. It intends to proceed along a number of fronts: collect basic environmental information about a number of Marine Corps programs, identify current and emerging families of technologies that might be used in C2 equipment for such programs, perform modeling and analysis to derive environmental guidelines and/or requirements for the component technologies of such equipment, and work with manufacturers to persuade them to adjust their equipment so that it better meets the needs of the Marine Corps.

Findings

The basic premise of this program is unsound, and the specific tasks it intends to accomplish are dubious. The basic premise appears to be that the Marines can perform a survey of "families of commercial technology," maintain the information so that it is reasonably current, and use the results of this survey plus some program-based analysis to convince manufacturers of commercial equipment to adjust their products so that they better meet Marine Corps requirements for ruggedness, operability temperature ranges, and so forth. The committee had a very difficult time in learning exactly what work was proposed. The one concrete example given was that the program might research hard drives to discover the lowest temperature at which a drive could operate and then, by modeling, determine whether this bound could be improved (i.e., lowered further) by using a different kind of lubricating oil. The program's staff then hoped to approach hard-drive manufacturers with this information and convince them to change the oils used in their disk drives so that the disks would be better suited for use by the Marines.

One weakness in this approach is that commercial vendors are exceedingly unlikely to redesign their products so that they better suit the Marines. The Marines buy an extremely small quantity of equipment as judged by commercial standards, and the design efforts of commercial firms will aim to best meet their own marketing and cost-point goals. Ruggedization to very low temperatures, to salty air, and so forth is unlikely to rate highly on this list. Another weakness is the programs's lack of acknowledgment that specific types of equipment are introduced and replaced very rapidly within the commercial sector; it seems highly unlikely that the program could adequately track the ever-changing offerings in commercial technology. For instance, a given hard-drive manufacturer's 10-GB offering may use quite different components from month to month; thus experimental drop-tests of the equipment, for example, would be likely to give different results depending on the exact shipment. Finally, and most important, such detailed testing by the Marines is totally beside the point in today's performance-based acquisition environment. It is the contractor's responsibility to deliver equipment that meets a given program's requirements or to respond (as in the case of cost-as-an-independent-variable procurements) with suggestions for trade-offs between requirements and price. Thus in the committee's view, this program is entirely unnecessary and unlikely to be of any use.

Recommendations

The committee recommends that this activity be terminated immediately. It further recommends that no related effort be started in its place; this type of work is simply not needed.

Joint Tactical Radio System Mobile Network Radios

This program aims to ensure that the Marine Corps is fully involved in the development of the Joint Tactical Radio System (JTRS) and, in particular, its wideband ad hoc networking technology. JTRS will provide joint interoperability in two ways: by providing firmware to allow interoperation with radios that use legacy waveforms and by providing a new wideband network protocol for use by all the Services. The Marine Corps rightly views the wideband network as an essential enabling system for many of the new styles of expeditionary warfare being embraced. At present, the Marines are experimenting with prototype versions of such networks in the ELB ACTD by using existing VRC-99 and WaveLan data radios. These existing radios and the ELB prototype network are expected to be replaced by the JTRS wideband network as it becomes available. Thus it is essential that Marine Corps needs be represented as the JTRS network is procured.

Under this program, the SPAWAR Systems Center, San Diego, is chairing the JTRS radio-frequency (RF) networking integrated product team (IPT). The IPT is a rather large, joint, government team; at this point it includes tri-Service and joint Service, Department of Defense (DOD), and laboratory membership.[1] The essential goals of this IPT are twofold: to create a modeling and simulation environment in which Service concepts and scenarios can be tested with proposed networking approaches, and to develop a functional description document and a request for proposal for the JTRS wideband network waveform. The simulations are planned to begin almost immediately, with the RFP to be issued in December 2000.

Findings

The committee applauds the briefing that it received on this subject, as well as the underlying concept of the JTRS mobile network and its constituent software radios. As highlighted in the ELB experiments, the Marine Corps has a fundamental need for robust data networking to provide an infrastructure for most or all of its network-centric operational concepts. If all goes well, the JTRS mobile network will satisfy this need. However, the committee believes that the program is facing significant challenges and offers both observations and recommendations vis-à-vis these challenges.

Requirements Creep and Cost Growth. The overall JTRS program may already be experiencing requirements creep and cost growth that, if unchecked, can lead to a well-known vicious circle: equipment costs grow steadily, thus reducing deployment of equipment, which makes it even more costly on a per-unit basis, and so forth into a programmatic death spiral. The JTRS is currently required to support 43 waveforms, only one of which is a new, wideband, data waveform suitable for mobile networks. The others are all legacy waveforms, generally for voice modes of older military radios. Thus JTRS already suffers from requirements creep. This is probably not the end, however. As the various Services begin to draw up their requirements for mobile networking, there will likely be a tendency to mandate the sum of all requirements as the minimal requirements for the new wideband waveform. The committee notes the obvious fact that a larger list of requirements will generally result in a more costly system, not to mention schedule slippage, and that this trend should be avoided if possible.

Divergence from Commercial Practice. The JTRS has no obvious analogs in the commercial world, which may indicate that the military approach to resolving mismatched radios may differ strikingly from that in the commercial world. As a simple example, there is no widely available commercial radio that

---

[1] At present this includes the Air Force Communications Agency (AFCA); Electronic System Command (ESC); Air Force Research Laboratory (AFRL); Training and Doctrine Command (TRADOC), Army; Communications Electronics Command (CECOM), Army; Marine Corps Systems Command (MARCORSYSCOM); Marine Corps Combat Development Command (MCCDC); Space and Naval Warfare Systems Command (SPAWAR); Naval Air Systems Command (NAVAIR); Deputy Chief of Naval Operations (Space, Information Warfare, and Command and Control), (N6); SPAWAR Systems Center, San Diego (SSC-SD), Extended Littoral Battlefield (ELB) Program Office; Defense Advanced Research Projects Agency (DARPA); Director for Command, Control, Communications, and Computers Systems Directorate (J6); Joint Interoperability Test Command (JITC); Naval Research Laboratory (NRL); Office of Naval Research (ONR); U.S. Joint Forces Command (USJFCOM); and the Joint Tactical Radio System (JTRS) Joint Program Office.

accepts a number of software modules (implementing waveforms) from different vendors. One cannot buy a cell phone and insert the Motorola software image for time division multiple access to replace the original Qualcomm image for code division multiple access, or indeed insert Sony software to turn it into an FM radio. Although the JTRS manufacturers consortium is often portrayed as the leading edge of the commercial industry forum for software radios, some of the committee members are skeptical on this point and believe that a faster, cheaper, and more reliable means of ensuring inter-Service radio compatibility is simply to replace older radios with newer radios that implement a common waveform. Of course this process will have to be repeated every few years, just as in the commercial world, as the newer radios in turn become obsolete.

Maturity of the Networking Technology. Because mobile networking technology is not yet fully mature, great care should be taken in this area. In addition, other important areas—relating to leadership, decision making, and cognitive overload—are currently being neglected. The Army and the Marine Corps now have significant experience with mobile networks in the tactical internet, near-term digital radio, and ELB efforts. This experience has generally shown that the technology holds great promise but still has a number of technical problems that need resolution. Thus additional technical shakedown exercises will be needed before such networks are as reliable as they must be. Indeed, the current technology is sufficiently immature that more R&D likely is needed, although it could probably proceed in parallel with the procurement and shakedown of the initial mobile networks.

Overall Network Architecture (access points linked by a backbone). The overall *architecture* of the ELB network—not to be confused with its individual "box" components and their ensemble performance—has many good points and could be considered for adoption in the overall mobile networking architecture for JTRS. This architecture employs a backbone or long-haul network that provides wide-area connectivity across the battlespace, together with a number of access points that tie individual users and their hand-held devices into the overall network. In the ELB network, the backbone was formed by VRC-99 radios and the access networks by standards-compliant (IEEE 802.11) radios. The committee makes no special recommendation regarding those radios but observes that the decomposition of the overall battlespace network into these two subnetworks could lead to a better design than one in which a uniform set of radios is interconnected as a mesh.

Recommendations

The committee recommends that Marine Corps representatives to the JTRS RF networking IPT should make every effort to minimize the number of requirements imposed on the JTRS radio and its wideband waveform, so as to hold down the unit cost of these devices.

The committee further recommends that the ONR should fund an active R&D effort in ad hoc wireless networking in support of the Marine Corps need for sea-based, terrestrial, and UAV-based network nodes. This effort should include building a significant number of such nodes and performing detailed technical experimentation with real networks. The committee further notes that the ultra-wideband (UWB) radios discussed below might be a good platform for these initial efforts but that this networking research should be taken on in addition to current UWB radio programs.

JTRS Antenna (Helmet, Vest)

The Combat Wear Integrated program is intended to develop advanced RF antenna designs that can be incorporated into helmets and/or vests for individual Marines. A particular goal is to devise antennas that are suitable for the man-portable variant of the JTRS. This is a technical challenge because of the projected frequency range for such radios (2 MHz to above 2 GHz). This program is being undertaken by SPAWAR Systems Center, San Diego (SSC SD), in partnership with a Naval Postgraduate School (NPS) research team. NPS is performing much of the fundamental theoretical work, simulations, literature searches, and the like; SSC SD is advising, building prototypes, and performing tests. Radiation hazards will be analyzed by a theoretical homogenous model of the human body, validated by power measurements on a mannequin model of a human body.

Findings

The committee offers no particular comment on the technical feasibility or desirability of this program but is concerned about the potential RF/body interaction for such systems. In particular, public perception of potential physiological issues could be more important than any technical facts. There is already noticeable concern about potential health issues associated with the use of cell phones, and this concern could grow in coming years (whether or not it has medical justification). Thus, body-worn antennas could become a public relations problem; in the worst case, of course, they might actually pose a serious health issue for Marines who wear them. The committee is not convinced that the program's proposed experiments to measure RF emissions into the body have sufficient physiological realism to produce medically accurate results, but even if they did, it is an ongoing research task to determine how such results correlate with actual health effects on human users.

Recommendations

The ONR should continue funding this research but should also look for alternatives to body-worn antennas (e.g., antennas mounted on rifles).

Mobile Direction Finding for Tactical Signals Intelligence

This program contemplates a brief investigation into methods of automatic signal identification and antenna spatial referencing and the transition of selected methods into the team portable collection system early in FY01. This effort will be followed by a 3-year program for the development of a prototype PC-based system for mobile direction finding.

Findings

Reported past work involving calibration tests on the vehicle antenna combination of a proposed system seemed to fall more in the realm of system engineering/technical assistance than S&T. The committee was unable to discern whether the proposed future work was part of an acquisition program, an advanced technology demonstration, or exploratory. Reimplementing existing algorithms for a PC does not constitute S&T, nor does solving the system engineering problems of cooperative operation, but there are thorny S&T problems in automatic signal recognition and direction-finding arrays.

Some committee members think that there is ongoing work in automatic signal identification supported by the Defense Program 3 and the Combined Cryptologic Program and that the general problem is not platform-specific and not a good candidate for Service Program 6 R&D. Direction finding, however, is very platform-dependent because of the interaction of the platform, the antenna, and the environment.

Some on the committee are skeptical about the requirement for mobile direction finding by land vehicles. Because of the difficulties human operators face while traversing rough terrain, performance will be highly dependent on automatic signal recognition. A moving vehicle cannot erect a high antenna with line-of-sight to the target transmitter and therefore is condemned to operate at longer wavelengths whose near field encompasses the vehicle. Calibration to account for vehicle reradiation effects is very wavelength sensitive and may not always yield a unique actual direction of arrival for a given apparent direction of arrival.

Recommendations

The committee recommends that ONR should consult the National Security Agency, the Naval Security Group, and, particularly, the Naval Information Warfare Activity to determine the state of the art in automatic signal identification technology suitable for incorporation into lightweight ground equipment.

Marine Corps S&T investment should focus on direction-finding arrays suitable for vehicle mounting. The Army also has an interest in this problem. Demonstration of a suitable array/vehicle combination and revalidation of the requirement for direction finding on the move should be prerequisites for system prototyping, and the legitimacy of conducting such prototyping with S&T funding should be examined.

PC-based Time Difference of Arrival

This program seeks to define a generalized, PC-based time difference of arrival (TDOA) precision geolocation system utilizing multiple platforms interconnected by the single-channel ground and airborne radio system (SINCGARS) or similar limited-bandwidth organic communications. According to the program milestone chart, processing and location techniques will have been selected by mid-FY00, when the 1-year development of a processing package will begin. Demonstrations, brassboard construction, and transition to the engineering and manufacturing development (EMD) phase are planned for subsequent years.

Findings

Cooperative TDOA among units interconnected by channels of limited bandwidth poses two challenges: ensuring that all units are measuring the same signal and extracting and precisely time-measuring some aspect or statistic of the signal. For signals of adequate duration, units can be commanded to characterize whatever signal is found at a commanded frequency; dealing with shorter signals depends either on the ability of autonomous units to share the same search strategies measuring the same signal simultaneously, or the provision of large amounts of accurately time-stamped, predetection spectrum.

To the best of the committee's knowledge, past attempts at developing systems to geolocate accurately a wide variety of communications signals under the constraint of limited interconnection bandwidth and short baselines have failed. Many signals do not have easily recognizable features and do not lend themselves readily to the precise time measurements required for accurate geolocation with short baselines. Therefore, if improvements in the performance of digital signal processing now enable this performance, demonstration of this fact is a worthy S&T goal, although one that is not specific to the Marine Corps. Information on the algorithm selected for implementation and the results of analyses or simulations demonstrating the performance of the selected algorithm against typical communications signals would have been helpful. Absent these demonstrations, the committee is concerned that the contractor may be implementating a scheme of limited applicability.

It should also be noted that TDOA systems are not immune to the antenna/vehicle interactions that plague direction finding. In both cases, reradiation from the vehicle perturbs the measurement.

Recommendations

The committee recommends that ONR review the evidence that led to the selection of the algorithm being implemented, and if it is dissatisfied with this evidence, either terminate the program or redirect it into a more scientific exploration of the ability of various processing algorithms to accurately time-stamp various communications signals, an exploration that can be conducted without prototyping real-time hardware.

Wideband Tactical Communications

This dual-use S&T program will produce 12 UWB prototype radios for experimentation by the Marines Corps. Total program funding is $2.88 million, of which $1.22 million is paid by industry. The program combines two distinct technologies—UWB radios and ad hoc networking—to produce a self-organizing RF network that promises to be low probability of intercept (LPI), low probability of detect (LPD), and antijam (AJ). The radio will consist of UWB components integrated into a conventional Racal multiband inter/intra team radio (MBITR) platform. Ad hoc networking software will be added. Hardware delivery is planned for June 2001.

Findings

The committee commends this program. The basic approach is technically sound and can be expected to provide useful functionality to the Marines. However, the committee did not discern any planned path by which this new technology could be injected into Marine Corps operations for experimentation. In addition, the mobile networking scheme adopted for this new network will likely fall into the R&D prototype category and hence should not be confused with a production network that will operate reliably.

Ultra-wideband radios are a "hot" area at the moment and, perhaps as a result, their benefits are sometimes overstated. They are widely claimed to be LPI, LPD, AJ, and immune to multipath problems. There is an element of truth to all these claims, but UWB radios are by no means a panacea. Here, the committee notes that this technology looks promising for a number of Marine Corps applications, perhaps most notably MOUT. The proper choice of an operational frequency band for wall-penetration, taken together with UWB resistance to multipath problems that characterize urban environments, could yield a very useful hand-held radio for the Marines. Perhaps the most intriguing benefit is nontechnical, namely that UWB radios can probably be added as overlays across spectrum that is already fully in use with different radio technologies. This would allow the Marines to employ new types of voice/data networks while still maintaining legacy radios for some time.

Despite the potential for the UWB networks being prototyped in this program to have direct applicability in MOUT scenarios, the committee did not see any signs that ONR or the Marines had determined exactly how they would use these radios once the network was in hand. The committee suggests, therefore, that ONR and the Marines work out a plan by which the network can be used in a series of MOUT and/or Ship-to-Objective Maneuver (STOM) experiments in the next few years. Perhaps the most immediately useful role would be as a relatively high-speed data radio in the Small Unit Logistics (SUL) ACTD, which is badly crippled by the need to send relatively large amounts of data over very slow tactical voice radios (SINCGARS). The SUL experiment might work much better if it employed these new UWB radios instead of SINCGARS.

There will be a trade-off between data rate and penetration of building walls. Tests in the MOUT applications should examine this issue. The committee understands that the Federal Communications Commission recently made a ruling with respect to UWB radios, and it is important to determine if that ruling applies to the DOD and, if so, what if any limitations would be imposed. The performance of UWB radios advertised by various companies should be analyzed—if it has not already been—to determine if a sample quantity of these radios should be included in some of the tests.

Some members of the committee expressed the opinion that the mobile ad hoc networking (MANET) protocols being ported into the UWB radio are not technically mature but are instead more at the R&D experimental stage. In the committee's opinion, they will probably work well enough for experiments with small numbers of network nodes (e.g., 10) and so should prove useful in limited Marine Corps experiments with this technology. They should not, however, be confused with mature technology, and care should be taken lest the ELB experience (see Chapter 8) be repeated. The network will not necessarily work as assumed, and experiments should not take its proper functioning for granted.

Recommendations

The committee recommends that ONR and the Marine Corps work together to map out the set of Marine Corps experiments that should employ the prototype UWB radios (network) created by this program. It further recommends that the SUL ACTD would be a good candidate.

Summary of Recommendations for Command and Control

A summary of recommendations for command and control is given in Table 6.1.

TABLE 6.1 Summary of Recommendations for Command and Control

Project	Recommendation
Harsh Environments	Terminate.
JTRS Mobile Network	Minimize JTRS requirements accretion.
	Explore ad hoc networking technology.
JTRS Antenna	Continue funding, but seek alternatives to body mounting.
Mobile Direction Finder	Coordinate with the National Security Agency and others.
	Focus on vehicle-mounted antenna performance.
	Reexamine funding category.
PC-based TDOA	Review feasibility; terminate or redirect to exploration of various processing algorithms.
Wideband Tactical Communications	Map out experiments.
	Consider for incorporation in SUL ACTD.

RECOMMENDATIONS FOR NEW PROGRAMS

Unique Marine Corps concepts of operation, such as Operational Maneuver From the Sea, motivate a number of promising new command, control, communications, computing, surveillance, and reconnaissance (C4ISR) research areas for ONR Code 353. In many of these areas, a modest amount of R&D can have a tangible and highly useful payback in the very near future.

The committee recommends that the following basic approach be adopted for identifying new C4ISR research areas for ONR Code 353:

- Start with planned Marine Corps concepts of operations such as OMFTS and MOUT.
- Given these concepts, solve Marine Corps-specific problems in C4ISR.

Although this prescription may sound like a platitude, it leads directly to several S&T areas that are highly Marine Corps-specific, technically challenging, and not too expensive. The following examples suggest promising new starts for Code 353.

Connectivity to Joint and National Sensors

A small force deployed far into a country would benefit greatly from the situational awareness that can be provided by the joint surveillance and target attack radar system, Guardrail, or other joint and national systems. In a network-centric world it would not be necessary for these forces to carry the actual receive terminals for such sensor systems. Instead, the Marines would link into the overall network and obtain information useful for their situation. The committee believes that Marine Corps access to the information provided by such nonorganic sensor systems would be of great benefit and urges Code 353 to tackle this problem.

Mobile Networking for Marine-Specific Problems

Although many organizations are currently funding research in mobile ad hoc networking, Marine Corps use of such networks is sufficiently unique that specific research in this area is advisable. Most research in ad hoc networks is focused on the Army problem—i.e., very large numbers of vehicles in a relatively dense mass, organized hierarchically into subcommands that occupy abutting geographic areas—which requires technical solutions for accommodating a density of network nodes rather than addressing other needs such as stealth, long battery life, and so forth. A number of Marine Corps operational concepts pose technical challenges that may be Marine Corps-specific, such as deep forward observers, a long line of ocean-going AAAVs following a mine-cleared lane, small tactical UAVs for both sensor emplacement and communications relays, and so forth.

Network Security Issues for MOUT and OFMTS

Marine Corps concepts of small teams that are deeply emplaced into cities or nonfriendly territory lead directly to problems in network security. In particular, enemy capture of one or more Marine Corps network nodes (e.g., manpack terminals) amounts to having an enemy inside the Marine Corps network. Such insiders can launch very damaging information attacks before they are detected and excised from the network. Of course, all the Services are subject to enemy capture and insider attacks, but the Marine Corps faces these challenges to an unusual and probably critical degree. Hence an active program for identifying these risks and helping to mitigate them would be extremely desirable. The Marines could probably gain much leverage by coordinating with programs in the other Services and in the Defense Advanced Research Projects Agency that are tackling similar problems.

Networking the Advanced Amphibious Assault Vehicle

The AAAV is a very important vehicle for the Marines because it directly supports Marine Corps attacks from sea onto land. The committee has repeatedly heard the broader, and more revolutionary, view from the Marine Corps that AAAVs can be used in very new styles of sea-based attack if (and only if) the commanders and forces in these vehicles can be fully networked while they are maneuvering from ship to shore. Such networking would give the commanders sufficient situational awareness to make informed decisions on the landing points and would allow them to communicate these commands to their forces. The committee notes that it is by no means evident that the AAAVs, as procured, will have the communications capability to allow such networking. There are a number of technical challenges involved in providing wideband networking capability to vehicles that ride low in the water, that pitch about, and so forth. The Special Forces are familiar with these problems. In addition, there are technical challenges in the logical transition from being attached to the Navy's shipboard networks at the start of an operation, to transitioning to the "en route" network as the AAAVs maneuver to shore, and finally to linking up with Army and other networks as other Services come ashore after the Marines. The committee encourages Code 353 to become involved with the AAAV program and support it via C4ISR programs that will specifically enable networked operation of AAAVs.

Considering Deception and Other Information Operations

Because informed maneuver involves maneuvering where there is no large concentration of adversary forces, and because perfect situational awareness cannot remove these forces from the vicinity of the objective, deception about the location and intent of Marine Corps forces seems to be an essential component of OMFTS and STOM. Although the committee recognizes the possibility that technologies for deception and other forms of information operations are being pursued elsewhere, it recommends that Code 353 assess whether others are paying adequate attention to technology that addresses Marine Corps needs. At a minimum, Code 353 should pay attention to the signatures of maneuvering Marine Corps forces and to technologies for producing decoys that reproduce these signatures.

Summary of Recommendations for New Investments in Command and Control

A summary of recommendations for new investments in command and control is given in Box 6.1.

Box 6.1 Summary of Recommendations for New Investments in Command and Control

- Connectivity to joint and national sensors
- Mobile networking for Marine-specific problems
- Network security issues for MOUT and OMFTS
- Networking the AAAV
- Deception and other information operations

CONCLUDING REMARKS

Timely and assured C4I has become critically important for the Marine Corps, but it is not clear to the committee that ONR Code 353's programs properly reflect this fact. The committee believes that the Marine Corps can, to a large extent, leverage the C4I systems being designed and built by other Services. This is, of course, far preferable to designing such systems for the sole use of the Marine Corps. However, its unique mission and operational concepts will lead to certain gaps where the Marines Corps must take the lead. Examples include providing assured, coherent C4I services to Marines in transit across the sea-land boundary in AAAVs and to Marines who have penetrated deeply into enemy territory. ONR Code 353 can be very effective in helping design economical and effective means to bridge these gaps.

The existing C2 programs are rather diverse. Indeed some of them may be serving no useful purpose. The coming year, therefore, may well be a good time for ONR Code 353 to take a systematic look at its programs and prune away those programs that are not closely focused on core needs for the Marines. It should devote all of its time and energy to filling in those key components of C4I technology and other aspects of information superiority that will not otherwise be funded and that will, if provided, directly enable the Marines' operational vision.

7

Basic Research (6.1)

OVERVIEW

The Office of Naval Research (ONR) Code 353 has developed a new 6.1 research program intended to focus primarily on Marine Corps needs. This new effort is funded totally with ONR Navy "blue" dollars (as distinguished from Marine "green" dollars) and, although the initial funding is small, the committee commends ONR for initiating this program and for involving the Marine Corps Combat Development Command (MCCDC) in the program planning and selection process. ONR plans to invest $1 million in the current year, to double funding in FY01, and to consider further increases in the out years.

Code 353 created a list of possible topics and sent this list to MCCDC, which added topics, prioritized the list to favor "general areas in which operational concepts in the 2020 time frame are likely to be most dependent on new technologies," circulated the new list to ONR scientific officers, and solicited leads to performers and proposals. Table 7.1 records the list along with the number of awards in each area. Unfortunately, there seems to be very little correlation between MCCDC priorities and the number of awards made.[1]

PROGRAMS REVIEWED

Ultra-wideband Radio Ranging Studies

These studies aim to design a ranging algorithm for ultra-wideband (UWB) radio systems. In particular, the algorithm will attempt to minimize false locks on noise and on multipath signals that have

---

[1] All of the awards are close to $95,000 per year except the three in the middle category involving lightweight power sources. Those three are funded at roughly $150,000 per year (the extra $50,000 will cover some experimental devices that have to be built).

TABLE 7.1 MCCDC Priority Areas and Number of Awards

MCCDC Priority	Topic	Number of Awards
Most Marine Corps-unique, and most clearly linked to future operations	Communications in MOUT environment	0
	Multisensor information integration	1
	Artificial intelligence pattern recognition compression	1
Interesting but lower operational payoff	Lightweight power sources	3
	Artificial intelligence modeling (autonomous, cooperative)	0
	Disease prevention	0
	Materials for coatings	0
Important but not very Marine-specific	Ultra-wideband/low-probability-of-intercept communication	3
	Broadband laser eye protection	0
	Uncooled lightweight thermal sensors	0

not propagated directly from transmitter to receiver and then to determine the time of arrival of the direct path signal as accurately as possible. A propagation measurement effort in support of this objective will develop a database of measured signals that have propagated outdoors over ranges that can be supported by the allowed UWB transmitter power, as regulated by the Federal Communications Commission (FCC). This database will also include propagation conditions, information on direct path obstructions, and so forth, and will have general applicability to communications designs as well as ranging algorithm development.

Findings

The committee believes that the work on UWB ranging represents achievable technology and hardly qualifies as basic research. It entails buying some wideband radios and creating pulses short enough that multipath problems can be resolved. This is very similar to the analysis and experimentation done by telephone companies before they install a new tower, especially in an urban environment. If the intent of this work is to specifically address urban environments, it should be recognized that the Defense Advanced Research Projects Agency (DARPA) and other agencies have already addressed data collection, modeling, and simulation of ultra-wideband communications in such environments. Moreover, many high-fidelity models exist for wideband communications and signal propagation. And finally, based on the small amount of information provided, the committee cannot help but wonder what this project will contribute and how it relates to the existing legacy.

Recommendations

The committee recommends that this project should be considered for the core 6.2 discovery technology program and the funds freed up for more fundamental research. If the project is funded as a 6.2 effort, then care should be taken to identify what measurements will be made, whether these measurements have already been made by other agencies, and what models will be developed.

Channel Coding and Estimation for Ultra-Wideband Impulse Radios

Assessing the viability of ultra-wideband communications for military applications requires a framework for accurately analyzing system performance. This proposed research addresses the need for a tractable, outdoor UWB channel model that will serve as the foundation for performance evaluation and examination of the various design trade-offs that exist at the physical layer. The other major thrusts of the proposed research include channel estimation and error control coding, which are essential to the successful realization and implementation of UWB communication systems. Because of the multipath resolution capabilities of such systems, numerous resolved signal components will have to be selectively combined at the Rake receiver in order to acquire enough signal energy for reliable communications. Such processing and channel estimation will be very complex. Powerful channel coding such as turbo codes can be applied to UWB communication systems to alleviate the burden on diversity-combining schemes by reducing the required received signal-to-noise ratio.

Findings

The findings and observations presented for the ranging studies in the preceding "Findings" section (page 56) apply equally to this project.

Recommendation

This project should be considered for the core 6.2 technology program.

Low-Power Complementary Metal-Oxide Semiconductor Implementation of Ultra-wideband Radios

This research project intends to investigate the implementation of UWB radios for short-range data transmission using conventional complementary metal-oxide semiconductor (CMOS) technology. The FCC's decision to allow UWB transmissions opens the door for investigations into the design of such radios, with the ultimate goal of a single chip realization. Such implementation will allow comparison of UWB transmission with more conventional forms of radio transmission, so that the trade-offs can be more clearly understood. The research areas that will be investigated on this grant include the efficient generation of ultra-wideband signals and their subsequent amplification for transmission, as well as the complementary issues of amplification of the received wideband signal and circuitry required for synchronization. In particular, an approach for generation of these signals that exploits the ultrafast transitions of deep submicron CMOS logic will be explored. A related research focus will be a UWB antenna design methodology that includes simultaneous optimization of the design of the antenna and CMOS circuitry (either low-noise amplifier or power amplifier).

Findings

The committee and the project director agree that this is an engineering task. Given the existence of extensive work on implementation of UWB radios, the committee wonders what specific contribution will be made by this investigation and what the associated design or implementation issues are. Even as a 6.3 task, this undertaking would be much more compelling if it had an industrial component.

Recommendation

This task should be considered for 6.3 funding.

The Information Theory for Optimal Aim Point Selection via Multiple Sensors

In the 21st century, systems for aim point selection and automatic target recognition will involve multiple sensors. This research will investigate the derivation of algorithm-independent performance metrics for aim point selection using information from multiple sources in clutter. The ultimate goal is to achieve the best aim point selection for the least resource allocation. Performance bounds for sensors give guidelines for such sensors. These bounds will be based on the Shannon theory of imaging science, calculating algorithm-independent metrics that quantify information gain about the underlying targets (sources) through the multiple sensors (channels). Information loss will be calculated incorporating ground-based clutter models, quantifying performance degradation resulting from natural environments.

Findings

Although this appears to be a competent multisensor information integration project, the committee notes that many similar research projects exist throughout the Department of Defense and wonders whether this project addresses any Marine-specific issues not addressed elsewhere. This new research could perhaps make a useful contribution, but its value would depend on its focus: What candidate sensor types or characteristics are assumed? What assumptions will be made concerning sensor performance and utilization? What are the assumptions about the observing environment? Although a theoretical study could be useful, it should be based on fairly realistic possible observing situations and environments. If it is not, it might have little value. Such a project should also consider using advanced simulation tools to generate numerical results.

Recommendation

The committee recommends that this project continue as a 6.1 effort but that it be structured to address Marine Corps-specific needs.

Multisource Information Processing in Mobile Environments

This project includes three main tasks:

• Exploration of a new class of nonlinear wavelets, called minimum/maximum preserving wavelets, for compressing digital terrain elevation data (DTED). This work aims to extend the feature-based compression concept to develop more efficient and effective representations for DTED so that a user without large bandwidth or computing power can quickly access and query large, high-resolution databases remotely.
• Exploration of data-feature clustering approaches to enable timely information delivery, including access, update, and retrieval from large multimedia data repositories in mobile environments where bandwidth and computing power are constrained. Clustering approaches could allow developing new paradigms for data analysis that help bridge the gap between storage, access, and manipulation of digital data.

- Exploration of level set methods for automatic image registration of multichannel synthetic aperture radar (SAR) images. Level set methods could become a comparatively contrast-insensitive means of aligning multiple images containing common features.

Findings

The committee does not understand how these three interesting and related tasks will be integrated into a coherent research project and has questions about the individual tasks that would have required interaction with the investigators.

The compression of DTED data using nonlinear wavelets has potential value. However, the committee would have liked to examine how its utility (data compression factors, accuracy, computational requirements, and so on) will be evaluated for DTED, against which current methods for data compression the new method will be compared, and what data sets will constitute the baseline for comparison.

Data-feature clustering is a very broad (and relatively old) topic in data mining. What research is planned in this area? There are literally dozens of feature-clustering algorithms (decision trees, neural networks, cluster algorithms, and so on). The committee would have liked to examine the choice of methods to be explored, plans for the creation of new methods, and the selection of features to be investigated or used.

Automatic image registration for multichannel SAR is also a well-researched area. Much work remains, but the committee would have liked to verify the investigators' awareness of existing work in this area.

Recommendations

This project is potentially useful and appropriate. But more information would be needed on project plans to determine if the research will be productive or will only rehash existing research. The topics as described are not integrated into a coherent whole. Moreover, the title of the research is a bit misleading, since it has only limited connection to a mobile environment or to multisource information processing. The committee recommends that the project be reviewed by appropriate ONR Code 31 personnel who have knowledge of other work in this field.

High-Energy-Density, Rechargeable, Thin-Film Batteries for Marine Field Operations

This project includes two tasks:

- Polymer-clay nanocomposite materials will be prepared and evaluated for use as electrolytes in lithium polymer batteries. This work will entail the use of amorphous derivatives of polyethylene oxide intercalated into various clays to create the polymer-clay nanocomposite electrolyte. It is anticipated that this approach may produce a polymer electrolyte with significantly increased ionic conductivity while maintaining acceptable mechanical properties.
- Lithium polymer battery prototypes will be prepared and tested. The batteries will consist of interdigitated submicrometer layers of polymer electrolyte sandwiched between a micrometer-thick lithium metal anode and a three-layer oxide metal-oxide bidirectional cathode. This approach represents a strategy for increasing the surface-to-volume ratio of the active cathode material while maintaining ease of assembly. It is anticipated that this approach will enable larger lithium polymer battery electrodes to more closely match the high-capacity efficiency of microbattery electrodes.

Findings

New polymers for lithium solid electrolyte batteries have long been sought after by many research groups. The motivation is to develop new polymers that are conductive enough to be practical without sacrificing the mechanical rigidity that makes solid electrolytes attractive in the first place for their ease of manufacture in odd geometries and configurations that would make them man-wearable. Further, polyethylene oxide is among the most widely studied of the polymers applied to lithium batteries. In fact many academic, government, and industry groups are active in this field and receive funding from the Department of Energy, the National Aeronautics and Space Administration, the Army, the Navy, the Air Force, and the Defense Advanced Research Projects Agency. The scope of work just described includes new materials, electrochemistry, and battery construction and design.

Although this research fits into the 6.1 category, it would be useful to focus on new materials that might perform much better than polyethylene oxide.

The second task appears to involve the development of a new cell design that would have a central cathode and electrolyte and anode on either side to improve energy density. Presumably the cathode would have to be thicker in this design, but there would be only half as many of them relative to conventional cell designs, which have one cathode for every anode. The committee trusts that a trade-off has been made showing that fewer but thicker cathodes provide an overall energy density increase for a given cell.

State-of-the-art lithium polymer cells are already thin (100 μm) with high electrode surface-to-volume ratios; therefore, the committee questions how much improvement in energy density can be expected via a new cell design as opposed to new energetic electrode materials.

Testing battery prototypes is not a 6.1-level task. Assembling new cell designs and packing cells together as a battery is appropriate at the 6.2 or possibly the 6.3 level of development.

Recommendations

The committee recommends that the materials task be continued as a basic research project, taking into account the findings noted above, and that battery prototyping be transitioned into a different budget category.

Development of Fuel Cells for Direct Electrochemical Oxidation of Strategic Fuels

This research focuses on recent success in achieving stable power generation from hydrocarbon fuels via direct electrochemical oxidation in a solid oxide fuel cell employing a copper-ceria-yttria stabilized zirconia anode. Research will address the following key issues associated with expanding the applicability of such devices to common logistic fuels such as JP-8: (1) increased power densities, (2) long-time stability against coking, (3) tolerance to sulfur contamination from the fuels, (4) increased electrode mechanical strength, and (5) the need for interconnect materials that will not promote coking.

Findings

A number of past research studies on solid oxide fuel cells (SOFCs) have already demonstrated the ability to directly oxidize fuels such as methanol or light hydrocarbons, e.g., methane, without the need for a separate "reformer" to produce hydrogen and carbon dioxide from fuel and steam. This direct oxidation operation is usually referred to as internal reforming. The ability to perform this operation with fuels such as JP-8 would be of considerable benefit in terms of logistics and cost.

It is ambiguous, however, whether the five issues to be addressed with the proposed research have been identified specifically because of the selection of JP-8 as a fuel or if they represent areas of SOFC research generally.

To the extent that the five issues are directly associated with the use of JP-8, then a number of comments apply, including the fact that JP-8 will be little different from any other liquid hydrocarbon in terms of power or energy density. Of course, it will be superior to compressed gases because of the higher storage densities. With respect to stability against coking, the committee observes that a proper electrocatalyst should facilitate the complete oxidation of any hydrocarbon (methane, JP-8, diesel, and so forth) to carbon dioxide without deposition of soot.

The committee agrees that logistics fuels are indeed liable to contain various concentrations of sulfur and that a compatible fuel cell anode electrocatalyst will be necessary. SOFCs are already significantly more tolerant of sulfur than are other fuel cell types, e.g., those with a proton exchange membrane. In still another area, it is not clear why JP-8 would require greater electrode mechanical strength than any other hydrocarbon or even hydrogen.

Finally, it is not the lack of a better electrocatalyst (faster, sulfur tolerant, and so on) that inhibits the maturing of SOFC technology compared with polyethylene matrix cell technology. The greater challenge has often been in the mechanical integrity of cell stack materials over long operating times with several turn-down cycles (repetitive thermal shock) and in the response to external shock and vibration (ceramics can be very brittle).

Recommendation

This research work should be continued, but the investigators should look for transitions into the 6.2 core program. If sulfur tolerance is only partial, then the sensors described in Chapter 4 in the "Bulk Liquids" section (pages 29-30) for assessing host nation and captured fuels should be designed to detect sulfur contamination.

Modeling of Power Systems for Marines

This research focuses on the optimization of electrical power systems and loads for future Marine Corps application. It includes development of modeling concepts suitable for the increasingly complex electrical systems of future dismounted soldiers and land vehicles. Modeling concepts will be explored that will be capable of ascertaining the total system consequences of technical advancement of the various system components, including both power generation and utilization. As such, the model will constitute a tool for optimizing current science and technology (S&T) investment toward meeting the future electrical needs of the Marine Corps during expeditionary operations.

Findings

This modeling and simulation task appears to lack basic research content. The presentation lacked clarity as to how the results from the Army's "Land Warrior" tests will be incorporated into the study.

TABLE 7.2 Summary of Recommendations for 6.1

Project	Recommendation
UWB Ranging	Consider for 6.2 funding.
UWB Channel Coding	Consider for 6.2 funding.
Low-power CMOS	Consider for 6.3 funding.
Multiple Sensors	Continue, but structure to account for Marine Corps needs.
Multisource Mobile	Program review by Code 31.
Thin-film Batteries	Continue materials task, but transition prototyping to 6.2.
Fuel Cells	Look for a transition to 6.2.
Modeling Power Systems	Look for an early transition to 6.2.

Recommendation

The committee recommends an early transition of this work into a 6.2 program.

Summary of Recommendations for 6.1

A summary of recommendations for 6.1 is given in Table 7.2.

RECOMMENDATIONS FOR NEW PROGRAMS

The committee suggests that the following basic research areas should command high priority in the FY01 and later 6.1 portfolios:

- Investigation of phenomena that could be incorporated into weapons of controllable lethality (see "Small Arms" under "Recommendations for New Programs" in Chapter 3, page 25),
- Other fundamental research that could lead to better devices and techniques for urban operations,
- Phenomena and devices for mine detection and countermeasures,
- Techniques for gaining and disseminating situational awareness in a Marine Corps context to enable informed maneuver, and
- Materials that could reduce the logistics burden of Operational Maneuver From the Sea (OMFTS).

With respect to the top two items on this list, it appears to the committee that MOUT should be a key basic research (6.1) goal, but unfortunately it seems to be poorly supported in the current program.

With respect to mine detection, the committee perceives a need to conduct some solid basic research (or at least some early technology studies) to establish the phenomenology that will help determine what systems will or will not work. Right now, in the Joint Mine Detection Technology program (see "Joint Mine Detection Technology" in Chapter 2, page 15), ONR seems to be adopting an "Edisonian" approach of trying this and trying that until something eventually seems to work. ONR needs to examine the real physics of the problem and the different environmental conditions and then try to find some good discriminating factors that will allow the researcher to do pattern recognition studies, false alarm investigations, and so on. Without this fundamental approach, there is very little prospect for a

BASIC RESEARCH (6.1)   63

> **Box 7.1 Summary of Recommendations for New Investments in 6.1**
>
> - Investigation of other phenomena that could be incorporated into weapons of controllable lethality
> - Other fundamental research that could lead to better devices and techniques for urban warfare
> - Phenomena and devices for detecting mines, particularly remotely, and for mine countermeasures
> - Techniques for gaining and disseminating situational awareness in a Marine Corps context to enable informed maneuver
> - Materials that could reduce the logistics burden of OMFTS

significant advance in this important warfare area. Accordingly, the committee recommends that ONR Code 353 set up a sustained, coordinated 6.1 program to establish the scientific basis for the detection of mines in the littorals, particularly remote detection, and for mine countermeasures.

Situational awareness is obviously an important warfare capability for the Marines, but it appears to the committee that most of the problems and issues can be addressed with more investment dollars for equipment and software rather than basic research. Some of the more important issues include the exchange of tactical targeting data between forces ashore and those afloat; common data formats, update rates, and interface protocols for interoperability with other Services; and common displays. Even so, the committee recommends that the proposed research planning team examine the prerequisites for informed maneuver, including data fusion and dealing with information overload, to see if a good basic research program can be developed. The need to transform information into knowledge and the need for precise information in urban terrain, both called for in Chapter 2's overview (pages 14-15), provide a formidable 6.1 challenge.

Finally, the committee notes that the logistics burden will likely limit the feasible scale of OMFTS and suggests that a Marine Corps-oriented 6.1 program should include a search for higher-strength, lower-weight materials, and, perhaps, for fuels, propellants, and explosives of higher specific energy.

Summary of Recommendations for New Investments in 6.1

A summary of recommendations for new investments in 6.1 is given in Box 7.1.

CONCLUDING REMARKS

Although it commends ONR and its Code 353 for initiating a 6.1 program directed toward research with possible Marine Corps applicability and for involving MCCDC in project selection, the committee found that the content of the initial program was not compelling with respect to scientific merit or Marine Corps specificity. It notes the poor correlation between MCCDC's leading priorities (see Table 7.1) and the projects selected.

The committee recognizes the difficulty of identifying tasks that are 6.1 in nature and also have clear Marine Corps relevance; it also recognizes that the calendar limited the time available for making the first year's selections and that MCCDC, although an ultimate authority on what future capabilities the Marine Corps needs, is not particularly experienced in basic research program development. The committee realizes that its lack of direct interaction with the projects' principal investigators may have caused it to overlook some of the scientific content of the proposed work.

The committee urges the continuation and expansion of a Marine Corps-oriented 6.1 research program, but urges the following actions to help ensure a better outcome:

- Start the selection process sufficiently early that deadlines do not cause hasty actions.
- Institute a standing team that includes the ONR, MCCDC, Marine Corps Warfighting Laboratory, operators, and trainers to engage the Marines in program formulation more frequently than annually and to begin to demonstrate to the Marines that basic research can benefit the Corps.
- Engage the skills of ONR program officers and managers in departments other than Code 35 in assisting in the two-way translation between basic research goals and operational needs.

The third action applies to domains beyond 6.1 research; Chapter 9 (in "Activities of Code 353 Personnel," page 73, and "Other Desirable Activities," page 74) expands on this matter.

8

Extending the Littoral Battlespace (ELB) Advanced Concept Technology Demonstration

The Extending the Littoral Battlespace (ELB) advanced concept technology demonstration (ACTD) is a large-scale experiment with twin goals: (1) to demonstrate how the Marines will perform Operational Maneuver From the Sea (OMFTS)-type maneuvers when network-centric technology is available and mature, and (2) to integrate and refine existing technologies so as to provide support for such maneuvers. The ELB experiments involve ships at sea linked tightly in an information grid with Marines operating both dismounted and in vehicles at ranges of up to 200 nautical miles inland. Airborne relays are employed as needed to provide network communications over these large distances.

The committee understands that the Commander-in-Chief, Pacific Command has the following operational objectives for ELB:

1. Provide a battlespace communications network that is easily managed, scalable, and secure;
2. Provide a distributed command and control system with enhanced situational awareness and parallel planning;
3. Improve distribution of sensor and intelligence information;
4. Improve fire and targeting;
5. Provide security by encryption, network security, and intruder detection;
6. Demonstrate joint interoperability; and
7. Demonstrate force protection.

The first ELB ACTD employed VRC-99A radios for the wide-area ("long-haul" or "backbone") portion of the communications network, together with commercial WaveLan wireless local area network (LAN) technology for access networks. The VRC-99A radios were stationed on ships, aircraft, and ground vehicles and outfitted with antennas and power amplifiers that were sufficient to close the radio-frequency (RF) links at the expected ranges. These radios included ad hoc networking software and were expected to organize themselves into a backbone network that would carry traffic between the ships, aircraft, and ground positions. The radios were attached to WaveLan base stations so as to provide a standards-compliant (IEEE 802.11) wireless LAN access network that linked the command and control (C2) computers, hand-held computers, and so forth, into this all-encompassing network.

A number of technical issues were uncovered during the first ELB ACTD, but the operational results appeared promising despite the technical problems. Thus, a second ELB ACTD has been scheduled that will fix some of the obvious technical problems with the communications network and allow further operational experiments. This new experiment will also introduce newer C2 systems such as the integrated Marine multiagent command and control system, laptop servers down to the company level, multiple battlespace views, and interfaces to other (joint) systems such as Army tactical operations centers and the naval fires network.

The schedule features a number of shake-down experiments and tests that include FST-1 in June 2000 (focusing on communications and networking), FST-2 in conjunction with Millennium Dragon in September 2000, FST-3 in February 2001 (focusing on rehearsal), and the actual experiment, conducted in conjunction with Kernel Blitz (X) in June 2001.

FINDINGS

The committee believes that the ELB ACTD provides a striking mixture of good news and bad news. The good news is mainly operational—ELB has provided the first glimpse of how an OMFTS information infrastructure might work in practice. The bad news is entirely technical—the ELB network is an early prototype that is many years, and many tens of millions of dollars, away from a real production version of a tactical data network.

First, the good news:

- The RF links worked. The ACTD established link ranges in excess of 100 nautical miles with a 10-W power amplifier, had no trouble with Doppler shifts in air-to-air linking, and could deliver reasonable data rates (from 625 Kbps at 110 nautical miles up to 10 Mbps at 30 nautical miles).
- The multihop network showed promise. Network connections were established between El Centro, California, and ships at sea, and realistic applications were able to work across this network. The basic division of the network into a backbone and a set of access networks worked well.
- The experiment was the Marines' first real glimpse of how an OMFTS information infrastructure will work in a network-centric future. As such, it gave very valuable insights into the basic operational concept and demonstrated that, to a very large extent, the concept worked.

And now the bad news:

- The network suffered from a number of serious technical defects. It was often unstable and will likely have serious issues with scalability to larger numbers of network nodes (radios).
- The observed decrease in effective bandwidth as a function of range is a consequence of the decreasing signal-to-noise ratio as a function of range. Simply put, unless something can be done to maintain the effective system bandwidth, the system will only be able to support 6.25 percent as many users at a range of 110 nautical miles as can be supported at a range of 30 nautical miles.
- Applications and protocols designed for a highly stable campus LAN (e.g., an Ethernet) may tend to "commit suicide" in a tactical RF network. Congestion and changes in connectivity are much more severe than those encountered in most commercial environments, and a typical application program behaves very poorly under such circumstances.

- The straightforward application of commercial routing protocols to the highly unstable environment of a tactical RF network simply did not work—which was quite predictable given technical insight and the Army's prior experience in Task Force XXI. Under even a modest amount of user movement, the routing control traffic swamped the available network bandwidth; as a result, user traffic could not get to its intended destinations.
- There were both technical and operational challenges in trying to achieve an operationally useful apportionment of the available bandwidth between different classes of users. The network provided only a very limited form of quality-of-service control, and the higher-ranking users tended to seize all available capacity, e.g., for videoconferences.
- The system was highly vulnerable both to classic threats such as RF intercept and jamming as well as to new "network threats" such as enemy capture of a functioning network node.
- The highly dynamic tactical environment, in which participants enter and leave the network on a continual basis, also places unusual stresses on networking technology. In technical terms, users must be authenticated, their names and current locations must be recorded so that they can obtain services at their current attachment points, and so forth. Few of these problems are unique to tactical networks, but the overall rate of change in such networks generally exceeds that in the commercial arena and hence makes direct application of commercial techniques problematic.

The committee believes that the technical problems in ELB are deep-seated and fundamental. They will not be removed, or even significantly ameliorated, by a modest amount of tinkering. There is no quick fix here. Despite all wishes, tactical ad hoc networking is by no means a mature technology. In the committee's opinion, a very small, not-too-reliable ad hoc network of a dozen or so nodes is relatively easily achieved; that was what was demonstrated in ELB. However, it will take a *great deal* of time and money to build a production-quality network, i.e., one with adequate performance, scalability, and security.

Since network security is a critically important issue, and one that is usually poorly understood outside the technical community, the committee underlines its importance. Network security has little or nothing to do with RF link security (i.e., transmission security and communications security), though RF security will continue to remain at least as important as it ever has been. Network security involves protecting the Marines' tactical systems against network outages, corrupted databases, falsified calls for fire, and situational awareness that has been falsified by actions of an adversary. All these problems are classic information attacks—and they are highly likely if an enemy captures an ELB computer and attacks from "within." The committee has seen nothing discussed concerning the ELB network that protects against such information attacks. To repeat, this is a critically important area and must be addressed before any such network transitions to operational use.

RECOMMENDATIONS

Focus on Experimentation

The committee recommends that the ONR and the Marine Corps focus their efforts on the operational experimentation in ELB rather than on technical development. The goal should be to understand how operations may work once a production tactical data network is successfully deployed, rather than to try to deploy such a network.

Minimize Development Investments

The committee recommends that the radios and networks should be improved only to the *minimum extent necessary* for operational experimentation. It believes that money spent on improving the network or its communication links is unlikely to result in significant benefits because the network problems are deep-seated.

Learn How to Accommodate Outages

The ONR and the Marines should understand that network outages and glitches will be likely during the experiments no matter what and should design the operational plans and technical tests accordingly. The committee recommends that developing and testing procedures to deal with outrages be made an integral part of the ACTD.

Perform a Security Analysis

The committee recommends that ONR perform a security analysis of the ELB network and its communications links in order to determine and document its vulnerabilities. This analysis should be used as a basis for the design of the "production" tactical data network, e.g., in the Joint Tactical Radio System (JTRS) RF networking integrated product team (IPT).

Quantify Capacity Requirements

The committee recommends that the ONR use the ELB experiments together with modeling and simulation to quantify the amount and types of data that the Marines send across the network, together with the patterns of distributions of these data among different classes of users. This will provide useful (though very preliminary) information to the Marines and the JTRS RF networking IPT.

The ONR should attempt to quantify the ELB network's capacity in a realistic set of scenarios and ensure that such information is made available in a clear, easy-to-understand form to Marine Corps planners. The committee believes that it is unlikely that the operational users have any notion of how the network capacity varies with the distance between nodes (which can drastically affect link speeds), the amount of "churn" as users transition between wireless LANs, the number of active nodes in the network, and so on. A set of simple, quantitative guidelines may serve to avoid future problems caused by unrealistic user expectations.

Collaborate

Consistent with the recommendations of Chapter 6, the committee recommends that the ONR should work collaboratively with other agencies (e.g., the Defense Advanced Research Projects Agency and the Army Communications Electronics Command) to help develop mature ad hoc networking technology. In particular, Code 353 can readily and profitably begin work on ensuring that the Marines' special problems are addressed by ad hoc networking protocols. These problems include the networking of deep-forward observers, of advanced amphibious assault vehicles en route from ship to shore, and so forth, as discussed in Chapter 6.

TABLE 8.1 Summary of Recommendations for ELB

Project	Recommendation
NA	Focus on experimentation. Minimize development investment. Learn how to accommodate outages. Perform a security analysis. Quantify capacity requirements. Collaborate.

Summary of Recommendations for ELB

A summary of recommendations for ELB is given in Table 8.1.

CONCLUDING REMARKS

ELB is an important set of activities that has been allowing the Marine Corps a first chance to experiment with information infrastructures for its new operational concepts. As such, it is invaluable. The knowledge gained from these "test drives" will be extremely important as these new concepts are refined. There are two dangers inherent in all such experiments, however—first, that the "test rigs" themselves will be mistaken for fully operational systems, and second, that little by little more funds will be committed to improving the test rigs rather than to learning about what is required for an actual operational network. The ELB network and its components are the test rigs. The committee strongly urges ONR Code 353 to remain focused on the main task at hand—to gain insights and to start to gather quantitative data. ELB is a perfect opportunity to study and learn as much as possible in preparation for design of the real systems to follow.

9

Suggestions for Improving Program Effectiveness and Achieving Better Integration with the Marine Corps

This chapter provides suggestions for improving the management of the Office of Naval Research (ONR) Marine Corps Science and Technology (S&T) program to achieve better integration with Marine Corps operational objectives. As noted, the program inherited by Code 353 included many preacquisition activities and participation in large-scale demonstrations.

THE CODE 353 PROGRAM

Discovery Versus Demonstration and Experimentation

Finding

Given that the Marine Corps Warfighting Laboratory's (MCWL's) program is primarily experimental and that the future naval capability (FNC) programs will brassboard promising technologies for demonstration at MCWL and elsewhere, the committee believes that Code 353's program should be primarily a discovery program, that is, one that explores new technologies with an intent to clarify their applicability to Marine Corps needs. Although the Code 353 program need not exclude all demonstrations, those selected should feature promising techniques that do not fit in the FNC pipelines.

Recommendations

The strategy recommended for ONR that forms the basis of the committee's individual recommendations on the Code 353 program can be described as follows:

- Eliminate from the Code 353 program, at an orderly but determined pace, preacquisition and other activities that do not conform to the usual ONR S&T standards of innovation and technical aggressiveness.[1]

• Leave system demonstrations principally to MCWL, fleet battle experiments, and the FNCs.

• Embark on a discovery program to identify and refine technologies that can have a substantial payoff in achieving Operational Maneuver From the Sea (OMFTS).

• Exploit the talents and insights of other ONR divisions in program formulation and performer selection.

Transitioning Code 353's program to a discovery program will be challenging in light of the program's history and the interests and capabilities of its traditional performers; however, such a transition is necessary to build a much more technically exciting program than currently exists.

Balancing the Five Imperatives and Venturing Outside the Box

Chapter 1 (in "Lack of Quantitative System Analyses," pages 11-12) noted that the five technology imperatives identified by the Marine Corps Combat Development Command (MCCDC) were a reasonable taxonomy for S&T investment but that the needs were interdependent, with improvement in one area potentially reducing the requirement for improvements in others.

Findings

The committee did not find a process in place for balancing investments in the technology imperatives in terms of potential contributions to Marine Corps operational effectiveness. Instead, approaches chosen and even levels of investment seemed to be determined by the performing organizations. Performers are expected to be advocates of their assigned mission, but as one aphorism reminds us, if one has only a hammer, every problem looks like a nail. Furthermore, institutions sometimes regard maintenance of a level of investment in their technology and approaches as a right that does not have to be justified in terms of potential operational payoff.

In Chapter 1 (in "Lack of Quantitative System Analyses," pages 11-12) the committee noted that the general lack of quantification of force size and Ship-to-Objective Maneuver distance in the OFMTS concept could make it difficult to assess the value of a technology-enabled performance improvement. However, ONR can assist in the clarification of the OMFTS concept by identifying technologies that can support increasing force size and distance and estimating availability dates for these technologies.

Recommendations

The committee recommends that Code 353 build the necessary relations with MCCDC and other organizations to permit assessment of the operational payoff of a successful technology investment before investments are made in that particular technology, and that Code 353 have the authority to

---

[1] ONR's mission is to maintain a close relationship with the research and development community to support long-range research, foster discovery, nurture future generations of researchers, produce new technologies that meet known naval requirements, and provide order-of-magnitude innovations in fields relevant to the future Navy and Marine Corps. See ONR description online at <http://www.onr.navy.mil/sci_tech/>.

redistribute funds among the existing imperatives, new fields of endeavors, and existing and new performing organizations.

Because the Marine Corps has no organizational component responsible for establishing S&T requirements—a function now performed by a working group that meets periodically—the committee recommends that Code 353 take an active role in the following:

- Urging quantitative analysis of the OMFTS concept and providing, if necessary, some of the support and technology for such analyses; and
- Creating technology roadmaps that predict the performance levels and availability dates of the systems that are based on technologies that Code 353 and others are exploring.

CODE 353 AND OTHER PARTS OF THE OFFICE OF NAVAL RESEARCH

Need for a Marine Corps Line Organization Within ONR's S&T Activity

Some may question whether a technology-oriented Code 353 will have a program sufficiently distinct from the rest of ONR to justify the organizational distinction. Emerging technologies often do not have labels clearly indicating their most valuable applications.

Finding

ONR should have within its S&T organization a Marine Corps nexus that not only executes its own programs of special interest to the Corps, but also strives to bring all of ONR's talent to bear on Marine Corps needs. The committee does not believe that a staff organization outside ONR's main line organization can, by itself, ensure the results. However, for Code 353 to succeed in this function, it needs to fill its current vacancies with people who have a background in and a commitment to the Marine Corps, as well as the technology management experience to interact as peers with other ONR scientific and program officers.

Recommendation

The committee recommends that Code 353 be adequately staffed to perform the functions of executing a discovery program of special interest to the Marine Corps as well as to leverage, wherever possible, the remainder of ONR's S&T program to the benefit of the Corps.

Location of Code 353 Personnel

Most organizations attempt to locate their personnel in contiguous spaces; ONR Code 353, however, has placed each of its program officers near other ONR personnel making S&T investments in allied fields.

Finding

The committee concurs with the location of individual Code 353 program officers in spaces adjacent to those of other organizations pursuing similar technologies. All Code 353 personnel share a common

Marine Corps background and interest; many are active-duty Marine Corps officers. They require less additional interaction with each other than with other ONR personnel.

Recommendations

The committee recommends that ONR Code 353 continue its current strategy of physically locating its program officers near colleagues in other departments.

However, although propinquity may be a necessary condition for successful informal interaction, it is far from a sufficient condition. Code 353 will need to make a special effort to achieve effective liaison. ONR management should demonstrate its commitment to supporting this interaction in two ways as follows:

- Providing additional staff to Code 353 so that program officers can interact effectively with their ONR peers and Code 353 can participate more forcefully in the management of the Marine Corps S&T program, and
- Communicating clearly that support of the Marine Corps is the responsibility of the entire Office of Naval Research.

Activities of Code 353 Personnel

The committee considered whether Code 353 had a special responsibility beyond the planning and execution of an S&T program with a particular relevance to Marine Corps needs.

Findings

The committee recognizes that Code 353 program officers have been physically located so that, in planning their programs and selecting their performers, they can have the benefit of the insights on S&T of ONR personnel from other divisions. However, co-location will not, by itself, result in meaningful interaction.

Moreover, the dollar magnitude of the program conducted by Code 353 is less than 2 percent of ONR's S&T program. Those Code 353 program officers who succeed in sensitizing their colleagues in other divisions to Marine Corps needs can leverage their colleagues' much larger investments to the benefit of the Corps.

The Marine Corps depends on the Army for developing most land warfare technologies. Although the committee did not identify all the mechanisms that higher echelons of the Marine Corps may use to influence Army investments, the committee observes that there are mechanisms for inter-Service coordination that involve working-level technology managers.

Recommendations

The committee recommends that Code 353 seek mechanisms, in addition to its personnel location strategy, to develop a technically robust discovery program oriented to Marine Corps needs.

The committee recommends that Code 353 program officers, in addition to managing their own programs, interact vigorously with ONR colleagues in other divisions who work in similar technologies, not only so that the Code 353 program officers can benefit from their colleagues' expertise, but also so that they can influence the goals of their colleagues' programs by communicating Marine Corps needs.

The committee also recommends that Code 353 personnel attend reviews of Army S&T programs, participate in the appropriate inter-Service technology coordination forums, and seek to leverage other government and industry sources of technology.

Other Desirable Activities

One means of communicating needs is by conversation; another is to introduce the relevant Marine and ONR officials to each other. Members of the committee were favorably impressed by the April 2000 3-day "Open Space Meeting" at which 50 Marine Corps technology users from MCWL, Marine Corps Systems Command, and elsewhere interacted with 90 Naval Research Laboratory (NRL) technologists to learn from each other about mutual opportunities and needs.

Findings

The committee believes that what apparently worked well with NRL could work well with NRL's parent organization, ONR, although NRL has the advantage of having gadgets to demonstrate. Group interactions need to be well planned and supported; the success of the NRL interaction was doubtless a result in part of the diligent work of the Marine Corps contingent at NRL and in part of the support of the concept by senior officials at NRL and at the Marine Corps Combat Development Command, Quantico, Virginia.

Recommendations

Code 353 should plan a major group interaction between ONR program officers, ONR-supported researchers, and Marine Corps technology users, and, if the plan appears sound, ONR senior management should ensure the participation of a wide spectrum of ONR program and scientific officers.

THE OPPORTUNITY

The criticisms of the Code 353 program throughout this report may dishearten some ONR readers, but the committee wants to point out that it recognizes the circumstances that led to the current program and also that many of these circumstances represent opportunities. The establishment of MCWL provided a venue for verifying the payoff of improved technology. The transfer of management of the S&T program to ONR provides a current opportunity to purify it of system engineering/technical assistance and preacquisition activities. The initiation of FNCs will provide an opportunity to refocus the Code 353 S&T program toward new and technologically exciting directions. Perhaps most significantly, the people of Code 353 have an opportunity to go beyond managing the official Marine Corps S&T program and to influence the much larger Navy S&T program so that it will be of greater benefit to the Corps.

Appendixes

A

Terms of Reference

In response to a request from the Office of Naval Research, the National Research Council established the Committee for the Review of ONR's Marine Corps Science and Technology Program, under the auspices of the Naval Studies Board, to conduct a study as follows:

A review of the Office of Naval Research's Marine Corps science and technology research program will be conducted. The review will evaluate program components (in the areas of maneuver, firepower, logistics, command and control, and training and education) against criteria which the review committee will select such as: appropriateness of the investment strategy within the context of Marine Corps priorities and requirements; impact on and relevance to Marine Corps needs; Navy/Marine Corps program integration effectiveness; and scientific and technical quality. The review also will seek to identify promising basic (6.1), exploratory (6.2), and advanced (6.3) research topics that could be initiated to support the Marine Corps science and technology program.

B

Previous Training and Education Studies

During the 1990s, three major efforts were made by National Research Council committees to examine the role of science and technology in enhancing job performance and increasing human efficiency in the military services: the Navy Carrier-21 study,[1] the Army Star-21 study,[2] and the Technology for Future Naval Forces (TFNF) study.[3] Relevant findings and recommendations from these studies are reproduced here.

NAVY CARRIER-21 STUDY (1991)

The Carrier-21 study was the first of the three studies to deal with training and educational technology. Although the analysis focused on identifying emerging technologies that would affect future training aboard carriers, many of the findings and recommendations are relevant to on-shore jobs and units and directly to Marine Corps forces.

The Human Factors Technology Group of the Carrier-21 study stated that dependence on computer-based technologies would become as commonplace and essential as today's dependence on the telephone. Further, these technologies must be harnessed in the training arena in ways that would make training more accessible, convenient, relevant, inexpensive, and effective. The most important of these technologies and their relationship to critical functions is shown in Figure B.1.

---

[1] Naval Studies Board, National Research Council. 1991. "The Task Group Report of the Human Factors Technology Group," *Future Aircraft Carrier Technology, Vol. II: Task Group Reports (U),* National Academy Press, Washington, D.C., pp. 445-568 (classified).

[2] Board on Army Science and Technology, National Research Council. 1994. *STAR-21: Personnel Systems. Strategic Technologies for the Army of the Twenty-first Century*, National Academy Press, Washington, D.C.

[3] Naval Studies Board, National Research Council. 1997. *Technology for the United States Navy and Marine Corps: 2000-2035, Becoming a 21st-Century Force, Vol. 4: Human Resources*, National Academy Press, Washington, D.C.

APPENDIX B

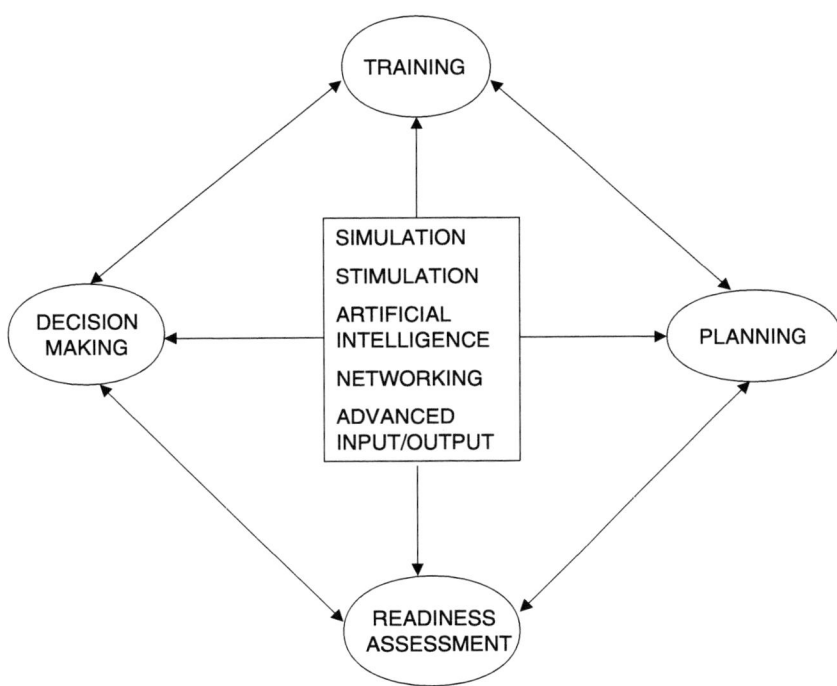

FIGURE B.1 Technologies and processes required for fleet readiness training. SOURCE: "The Task Group Report of the Human Factors Technology Group," *Future Aircraft Carrier Technology, Vol. II: Task Group Reports (U)*, National Academy Press, Washington, D.C., p. 527 (classified).

The group made the following recommendations to guide advanced simulation and training capabilities:

• Use available resources and procedures to speed the transfer of proven computer-based simulation and stimulation capabilities. As these technologies become more reliable and less expensive, they will become increasingly important.

• Move with deliberate speed to establish and use a standardized network of shipboard and shore-based trainers. A service standard for network linkages should be established and maintained. Both simulation and stimulation systems should be incorporated into a single system. Systems that will allow scenario generation, modification of training exercises to meet specific requirements, automatic recording and analysis of performance data, and timely feedback of readiness assessments must be overlaid on the integrated network to allow commanding officers and training personnel to tailor the training to the needs of specific crews and missions.

• Develop requirements, strategies, and facilities to support the development of a worldwide capability for task force and battle group training. This capability will be essential for allowing the battle force commander to conduct coordinated, mission-relevant training anytime and anywhere. A global network using satellite linkages will be critical for developing this capability.

ARMY STAR-21 STUDY (1994)

In the second of the three studies, one facet of the effort focused on personnel systems in the Army. The Personnel System Panel of the Star-21 study stated that training would become increasingly important. As missions became more unpredictable, the soldier would have to understand capabilities rather than doctrine and the job would become more strenuous. At the same time, rapid technological change would require continued retraining of experienced personnel. New jobs would place considerable demands on training technologies, which would have to:

- Be cost-effective and quickly applicable to emergent problems;
- Enable personnel to perform successfully in increasingly complex tasks;
- Be adaptable to different abilities and aptitudes;
- Be developed for specific social skills needed for teamwork roles; and
- Consider the motivation of the individual(s) being trained.

In addition to using cognitive science techniques to construct systems that are specifically intended for training, the Army should make extensive use of simulation systems—deployed in the field—to maintain unit readiness. The simulations should be user-reprogrammable to allow for a wide variety of computer systems that can be used to present meaningful tasks in forms that support efficient learning. The development of instructional tasks in these systems can utilize the expertise of experienced soldiers. Problems are presented with varying levels of help, so that students with different learning capacities can be trained in individually tailored sessions.

Not all training problems are solved by these suggestions. The available scientific base is deficient on three important topics. There is an insufficient understanding of (1) the learning that occurs on a job and the kind of technology that can make this learning more efficient, (2) how social interactions among workers promote or inhibit learning, and (3) the ways in which training prepares people to become effective learners and contributing members of a working group.

The panel concluded that to achieve better job performance and more efficient training, science and technology must be applied to the design of equipment for job training and to the management of the soldier's career. Table B.1 shows how areas of performance enhancement relate to one another, to technology, and to system management technology.

TABLE B.1 Potential Areas for Improvement in Army Personnel Systems

	Training	Technologies and Systems	System Management
Enhancing the performance of the individual soldier	• Develop research to find ways to improve physical and mental training • Second language • Intelligence gathering	• Optics, electronics, and computing • Visored helmet • Miniature computers and communications • Position-locating systems • Chemical/biological protection • Biomedical systems • Mobile mech. mastiff • Exoskeleton	• How much information is really needed by the soldier • Use model-based computer-aided design approaches • Practice user-centered design • Conduct mission-oriented simulations • Make use of psychometric techniques for the selection of battle managers
Human-machine systems (human factors engineering)	• Must train to work without advanced technologies (e.g., battle management system), in case of system failure or situation limitations (e.g., laser intercept capability (LIC)) • Develop technology to support predeployment mission rehearsal	• Sensor technologies • Computer assistance • Use reusable software • Develop analytical models for predicting human-system performance • Simulation-based training technologies for C3I	• Provide program support for research that is clear and enduring • Give high priority to research on the retention of skills
Selection, assignment, training, and career development (personnel systems)	• Support basic research aimed at a theory of situated learning • Examine theories of the transfer of knowledge from situation to situation • Study complex skills and their application • Study the development of teamwork skills • Study motivation • Conduct research on "learning to learn" • Cross-cultural training	• Examine technologies that promote on-the-job learning	• Augment present test batteries with computer-administered tests
The soldier as a system	• Behavioral and computer science research • Second language training	• Supersensory devices • Weapons systems • Personal information systems • Mechanical mastiff • Simulation technologies • Visored helmet • Lightweight chemical, biological, and radiological protective uniforms • Tailored vaccines	• Consider the soldier as a system

SOURCE: Board on Army Science and Technology, National Research Council. 1994. *STAR-21: Personnel Systems. Strategic Technologies for the Army of the Twenty-first Century*, National Academy Press, Washington, D.C., p. 4.

TECHNOLOGY FOR THE UNITED STATES NAVY AND MARINE CORPS, 2000-2035 STUDY (1997)

The most recent NRC analysis of technology applied to training was conducted by the Panel on Human Resources of the TFNF study in Volume 4. The overarching emphasis of the panel's analysis of technology as a means of increasing training effectiveness and efficiency was to "invest more in the conversion of conventional forms of training to technology-based, distributed training" (p. 8). The panel suggested that training capabilities being sought should be accessible, effective, and efficient.

Specific recommendations included the following (pp. 55-56):

"• Training should be accessible.
—Training should transcend physical location so that it is available wherever it is needed or wanted. Training should be available in schools, homes, workplaces, and learning centers. Environmental constraints should be minimal.
—Training should transcend time so that it is available whenever it is needed or wanted. Scheduling of training resources, equipment, materials, and/or instructors should not constrain the time at which training can be accessed.
—Training should transcend physical devices so that it can be portable. Constraints imposed by delivery platforms should be eliminated.
"• Training should be effective. It should do the right things. It must be relevant (1) to the job to be performed and (2) to the individual who is to perform it. Training analyses should be done in real time to set skill and knowledge objectives specifically tailored to the skills and knowledge that an individual needs.
"• Training should be efficient. It should do things right. Once relevant objectives are chosen, the instructional approaches used to meet them should be the most cost-effective available for the individual being trained."

The panel suggested that three areas of technological development seemed likely to change the nature of training: embedded training, modeling and simulation, and intelligent training systems. Technology-based training capabilities expected to be available by 2035 (or earlier) are listed in Table B.2.

Workplaces in all sectors are being continually infused with technology that requires workers to become increasingly technology literate. The Internet is quickly changing the way companies train their workers in technology. Internet-based lessons are replacing more traditional corporate classrooms, face-to-face meetings, and manuals. Technology is being used by companies to deliver current training on demand via the Internet to workers worldwide. Institutions of higher learning are also offering courses via distance learning. The panel, however, reported that in FY97 technology-based software was unlikely to be found in more than 4 percent of all Navy and Marine Corps training. It was suggested that the military could join with other federal agencies and the private sector to leverage the development of performance and certification standards for jobs and occupational areas of common interest.

APPENDIX B

TABLE B.2 Technology-based Training Capabilities Expected to Be Available in 2035

Capability	Description	Key Enabling Technologies	Goal
Embedded training	Training for operation, maintenance, and/or employment of a system (e.g., device, software package) included in, and presented by, the system itself	• Human-computer interaction • Information access and decision support technology • Cognitive modeling	• Obviate requirements for external training: potential user should need only to turn the system on to learn how to use it—all operator and deployment training should be embedded, as should most maintenance training • Ensure separation of training from operations and noninterference of one with the other
Distance learning	Structured learning that takes place without the physical presence of an instructor. Distance learning refers to distance training, distance education, distributed training, etc., and includes the full range of approaches (not just video teletraining) for distributing instruction to physically dispersed students	• Computer and video communications • Data compression • Networking • Interactive courseware (e.g., computer-based instruction, interactive multimedia instruction, techniques of individualization, design to effect specified outcomes)	• High-quality training available anytime, anywhere, to any student • Integration with personnel, classification, and assignment systems
Interactive courseware	Training delivered using computer capabilities that tailors itself to the needs of individual students	• Computer technology • Cognitive modeling • Instruction engineered to achieve specified training outcomes	Training that uses interactions with each student to maximize its efficiency by tailoring sequence, content, style, and difficulty of instruction to the needs of that student
Intelligent training systems	A form of interactive courseware that is generated in real time, is tailored to the needs of the individual student, and permits initiation of a tutorial dialogue and open-ended questioning by the student	• Speech and natural language interaction • Cognitive modeling • Knowledge representation • Computer technology	• An articulate, expert tutor for every student, possessing full knowledge of the student, the subject matter, and tutorial techniques and capable of sustaining mixed initiative, tutorial dialogue in terms the student understands • "An Aristotle for every Alexander"

continues

TABLE B.2 Continued

Capability	Description	Key Enabling Technologies	Goal
Simulation	Representations of real-world systems, situations, and environments that help achieve specified training objectives	• Digital, multimedia displays • Fidelity matched to training objectives • System, situation, and environment representation • Knowledge representation	• Device representations for maintenance and operator training generated directly from computer-aided design databases • Representations of interpersonal situations that respond to student decisions and actions • Representations of environments that convey sufficient psychological reality to achieve specified training objectives
Virtual reality	A form of virtual simulation—sensory immersing representations of real-world environments	• Digital, multimedia displays • Multisensory displays • Real-time interaction	Environmental representations providing full psychological reality and sufficient physical reality selected to achieve training outcomes
Engagement simulation	Simulations providing live, virtual, and constructive representations of real-world warfighting environments	• Networking • Data communications • Digital, multimedia displays	Seamlessly linked simulations supporting simulated environments in which engagements occur continuously against "real" and semiautomated forces
Human performance assessment	Assessment of relevant performance capabilities of individuals and teams	• Psychometrics of simulation • Job-sample testing • Assessment of cognitive processes	Valid (measures the right thing), reliable (measures things right), and precise (exactly identifies progress toward learning objectives) assessment of the knowledge, skills, and attitudes of individual students and teams available at any time in a training program

SOURCE: Reprinted from Naval Studies Board. National Research Council. 1997. *Technology for the United States Navy and Marine Corps: 2000-2035, Becoming a 21st-Century Force, Vol. 4: Human Resources,* National Academy Press, Washington, D.C., pp. 58-59.

C

Committee Biographies

Bruce Wald is an adjunct member of the research staff of the Center for Naval Analyses, as well as the founder of Arlington Education Consultants, an independent consulting organization serving both government and industry. Dr. Wald has an extensive background in electronic warfare, communications, space surveillance, and computer architecture, particularly in regard to their implications for naval and national security issues. Following college graduation at the age of 18, Dr. Wald joined the Naval Research Laboratory (NRL) in 1953, where he spent the next 33 years in positions of progressively increasing line responsibility in system and technology development, in project and group leadership, and in senior management. In his last position at NRL, Dr. Wald served as associate director of research and director of space and communications technology. Previously, he was superintendent of NRL's Communications Sciences Division, and before that the founding head of its Computer Science Branch. Dr. Wald has served on numerous panels of government advisory and scientific advisory boards, including the Army Science Board, the Defense Science Board, and the Naval Research Advisory Committee. Dr. Wald is currently a member of the NSB.

Alan Berman, an independent consultant, currently consults for the Center for Naval Analyses, where he assists with analyses of Navy R&D investment programs, space operation capabilities, and information operations. He also consults for the Applied Research Laboratory of Pennsylvania State University (ARL/PSU), where he provides general management support and program appraisal. Dr. Berman's background is in defense science and technology, particularly in regard to advanced weapon and combat systems. His previous positions include dean of the Rosenstiel School of Marine and Atmospheric Sciences at the University of Miami and director of research at the Naval Research Laboratory. Dr. Berman has served on numerous government advisory and scientific boards. He is currently a member of the NSB. He is also a member of the Free Electron Laser (FEL) oversight board that advises Jefferson National Laboratory of the Department of Energy on its FEL program.

A. Douglas Carmichael is professor of power engineering at the Massachusetts Institute of Technology (MIT). Dr. Carmichael's background is in naval propulsion systems. He joined the Department of Ocean Engineering at MIT in 1970, where submarine propulsion was his primary research interest. Prior to joining MIT, Dr. Carmichael was a research fellow at the Imperial College of Science and

Technology, London. He is the author of numerous publications on naval propulsion, including design impacts on alternative technologies. He is a fellow of the Society of Naval Architects and Marine Engineers.

Sabrina R. Edlow is research team leader of the Mine Warfare Systems Team at the Center for Naval Analyses (CNA), where she develops and executes the mine warfare program and other mine-warfare related projects. Ms. Edlow recently led an assessment of war plans and executed operations for *Desert Thunder* and *Desert Fox*. Her research interests encompass a wide-range of areas, including naval force structure planning, mine warfare, and underwater acoustical systems. A nuclear engineer by training, Ms. Edlow began her career as a design engineer at Duke Power Company, where she coordinated the nuclear fuel supply for seven nuclear reactors.

Brig "Chip" Elliott is principal scientist at BBN Technologies. Mr. Elliott's background is in Internet and wireless network technologies, tactical communications systems, and space-based surveillance and communications. As the technical lead scientist at BBN, he uses Internet technology to build networks for international corporations and U.S. government agencies. He was the chief architect for the networking component of the U.S. Army's Near-Term Digital Radio Program, which forms the backbone of the Army's Tactical Internet, for the British Army's High Capacity Data Radio network, and for the Canadian Army's IRIS network. He has also acted as lead for a number of low Earth orbit satellite systems (Discoverer II, SBIRS Low, Celestri) as well as a proposed undersea network.

Charles A. Fowler, an independent consultant, is retired senior vice president at Mitre Corporation, a federally funded research and development center serving the government on issues relating to national security. Mr. Fowler, a member of the NAE, has an extensive background in military systems, particularly those systems combining sensors, platforms, and command, control, and communications. Mr. Fowler began his career in 1942 as a staff member of the Radiation Laboratory at the Massachusetts Institute of Technology, where he participated in the development and testing of the GCA radar landing system. He later went on to engineering and management positions at the Raytheon Systems Company before joining Mitre in 1976. Mr. Fowler is a fellow of the American Institute of Aeronautics and Astronautics, as well as the Institute of Electrical and Electronics Engineers.

Ray "M" Franklin, a retired Major General, U.S. Marine Corps, is an independent consultant. General Franklin once headed the Marine Corps research and development effort. Today, he serves as an independent consultant on matters related mainly to amphibious warfare and force projection. He is particularly knowledgeable about research and development, systems acquisition, and military operations and training. A naval aviator, General Franklin has experience in both rotary and fixed-wing aircraft. He has participated in numerous government advisory and scientific boards, including that on littoral warfare for the Naval Research Advisory Committee.

David B. Kassing is associate corporate research manager for Defense Planning and Analysis at RAND. Mr. Kassing's background is in military logistics and deployment systems. He recently led an analysis of reception, staging, onward movement, and integration for the U.S. Army and directed a study of future materiel distribution systems for the Department of Defense. Mr. Kassing also reviewed the methodologies used by the Armed Services to set conventional munitions requirements and programs. He is a past president of the Center for Naval Analyses at the University of Rochester.

R. Kenneth Lobb is a senior scientist at the Applied Research Laboratory of the Pennsylvania State University (ARL/PSU). Dr. Lobb is an aeronautical engineer whose expertise includes aircraft design, advanced structures, navigation systems, underwater acoustics, underwater vehicles, and systems engineering. Over the course of his career, Dr. Lobb has served in a number of progressively increasing line responsibility positions, including aeronautical engineer at the Naval Ordnance Laboratory, technical director at the Naval Air Development Center of the Department of Defense, and vice president at the Center for Naval Analyses.

Irwin Mendelson is retired president of the Engineering Division of Pratt & Whitney. A mechanical engineer by training, Mr. Mendelson's background relates primarily to commercial and military aircraft engine design. As president of the Engineering Division at Pratt & Whitney, with a total staff of 8,000 and an annual budget of $900 million, Mr. Mendelson was responsible for the total operation of the division, including the design, development, and assembly of all aircraft engine systems. During his career, Mr. Mendelson has been directly responsible for the design and development of turbofan engines, jet engine fuel controls, pyrophoric ignition systems, and thrust controls for rockets.

Herbert Rabin is interim dean of the School of Engineering and director of the Engineering Research Center at the University of Maryland (UMD). A solid-state physicist, Dr. Rabin has an extensive background in applied technology; his research interests include quantum optics and space science. Prior to joining UMD, Dr. Rabin served in a variety of senior positions for the Department of the Navy, including deputy assistant secretary of the Navy for research applications and space technology. Dr. Rabin is a fellow of the American Physical Society.

David A. Richwine, a retired Major General, U.S. Marine Corps, is executive vice president of the Armed Forces Communications and Electronics Association (AFCEA), a nonprofit organization whose goals are to provide a forum for government and industry leaders to exchange ideas and concepts, discuss current problems and solutions, and identify future requirements in the technical disciplines of communications, electronics, intelligence, and information systems. General Richwine's military background is in C4I systems, primarily the interface of these systems with military operations. Before his retirement in 1997 from the Marine Corps, General Richwine served as assistant chief of staff for C4I and director of intelligence for headquarters.

Charles H. Sinex is logistics program manager for the Joint Warfare Analysis Department at the Applied Physics Laboratory of the Johns Hopkins University (APL/JHU). Dr. Sinex's background is in operations and systems analyses, with an emphasis on logistics and environmental issues relating to military operations. He recently led a major logistics effort for the deputy under secretary of defense for logistics on technologies for improvement of linkage models between military forces and logistics systems. Prior to this assignment at APL/JHU, Dr. Sinex served as supervisor of the Environmental Group, where he was responsible for numerous environmental survey design efforts.

Michael G. Sovereign is professor emeritus of command, control, and communications (C3) at the Naval Postgraduate School. Dr. Sovereign's background is in C3, joint warfare analysis, and acquisition cost-cycle analysis. He was formerly the director of the Institute of Joint Warfare and served as a visiting research professor for headquarters, U.S. Commander-in-Chief Pacific Fleet, where his responsibilities included conducting research on the Navy's Virtual Information Center workshops and other experiments aimed at addressing joint C4ISR issues. Dr. Sovereign was also senior principal scientist at SHAPE Technical Center (now NATO C3 Agency), where he participated in major re-planning of NATO C3 systems, and once served as director of special projects in the Office of the Assistant Secretary of Defense (Comptroller), where he directed the revision of DOD's planning, programming, and budgeting system and instituted methods for measuring and budgeting for inflation in weapon systems. Dr. Sovereign has authored numerous articles on instructional media, defense logistics, and economics.

Joseph Zeidner is professor emeritus of administrative sciences and psychology at the Graduate School of Arts and Sciences, George Washington University (GWU). Dr. Zeidner's background is in human resources and the factors involved in learning, training, decision making, and job performance. He has written several books on the economic benefits of predicting job performance and estimating the gains of alternative policies in affecting human performance. He has been influential in personnel classification issues and contributed to the *Encyclopedia of Human Intelligence*. Prior to joining GWU in 1982, Dr. Zeidner served as the technical director of the U.S. Army Research Institute and as chief psychologist of the U.S. Army.

D

Acronyms

AAAV	advanced amphibious assault vehicle
AAV	amphibious assault vehicle
ACTD	advanced concept technology demonstration
AJ	anti-jam
AMRAAM	advanced medium-range air-to-air missile
ASTAMIDS	Airborne Stand-off Mine Detection System (program)
ATD	advanced technology demonstration
AWT	Amphibious Warfare Technology (program)
C2	command and control
C4I	command, control, communications, computing, and intelligence
C4ISR	command, control, communications, computing, intelligence, surveillance, and reconnaissance
CB	current budget
CLAWS	Complementary Low-altitude Weapon System (program)
CMOS	complementary metal-oxide semiconductor
COBRA	coastal battlefield reconnaissance and analysis
COTS	commercial off-the-shelf
DARPA	Defense Advanced Research Projects Agency
DF	direction finder
DOD	Department of Defense
DTED	digital terrain elevation data
EFS	expeditionary fuel system
ELB	Extending the Littoral Battlespace (ACTD)
EMD	engineering and manufacturing development
ERGM	extended-range guided munition

ETAL	Enhanced Target Acquisition and Location (program)
FBE	fleet battle experiment
FCC	Federal Communications Commission
FNC	future naval capability
FOM	federated object model
FY	fiscal year
GPS	global positioning system
HMMWV	high-mobility, multipurpose wheeled vehicle
IEEE	Institute of Electrical and Electronics Engineers
IPT	integrated product team
JMDT	Joint Mine Detection Technology (program)
JTRS	Joint Tactical Radio System (program)
LAN	local area network
LASM	land-attack standard missile
LAV	light armored vehicle
LPD	low probability of detection
LPI	low probability of intercept
LRS	littoral remote sensing
MAGTF	Marine air-ground task force
MANET	mobile ad hoc networking
MARCORSYSCOM	Marine Corps Systems Command
MBITR	multiband inter/intra team radio
MCCDC	Marine Corps Combat Development Command
MCWL	Marine Corps Warfighting Laboratory
MEFF-V	MAGTF expeditionary family of fighting vehicles
MILES	multiple integrated laser engagement system
MOUT	Military Operations in Urban Terrain
MRRS	multi-role radar system
NA	not applicable
NASA	National Aeronautics and Space Administration
NPS	Naval Postgraduate School
NRC	National Research Council
NRL	Naval Research Laboratory
NSB	Naval Studies Board
NSWC	Naval Surface Warfare Center
OCSW	Objective Crew-served Weapon (program)
OMFTS	Operational Maneuver From the Sea
ONR	Office of Naval Research
OSD	Office of the Secretary of Defense

PEG	program evaluation group
POM	program objective memorandum
RF	radio frequency
RIS	range instrumentation system
RSEDG	Rapid Synthetic Environment Database Generation (program)
RST-V	reconnaissance, surveillance, and targeting vehicle
S&T	science and technology
SAR	synthetic aperture radar
SBA	simulation-based acquisition
SBIR	small business innovative research
SHORAD	short-range air defense system
SINCGARS	single-channel ground and airborne radio system
SOFC	solid oxide fuel cell
SPAWAR	Space and Naval Warfare Systems Command
SSC SD	SPAWAR Systems Center, San Diego
STO	shoot through obscuration
STOM	Ship-to-Objective Maneuver
SUL	Small Unit Logistics (program)
SUTT	Small Unit Tactical Training (program)
SYSCOM	systems command
TD	technical demonstration
TD1	first technical demonstrator
TDG	tactical decision game
TDMA	time division multiple access
TDOA	time difference of arrival
TFNF	Technology for Future Naval Forces (study)
TLDHS	target location, designation, and hand-off system
TLoDS	tactical logistics distribution system
UAV	unmanned aerial vehicle
UGV	unmanned ground vehicle
UWB	ultra-wideband
VE	virtual environment
VIRTE	virtual technologies and environment